GAS KINETICS

GAS KINETICS

GAS KINETICS

G. L. PRATT
The School of Molecular Sciences,
University of Sussex

JOHN WILEY & SONS LTD
LONDON   NEW YORK   SYDNEY   TORONTO

Library of Congress Catalog Card No. 76–80112

SBN 471 69640 4

Printed in Great Britain at the Pitman Press, Bath

CONTENTS

PREFACE

Recent years have seen a considerable change in emphasis among the topics conventionally included in chemical kinetics. Advances in our knowledge of the microscopic details of homogenous gas reactions have been particularly rapid. Studies of the details of molecular dynamics now occupy many kineticists. The reactions of atoms, free radicals, ions, and other excited molecular species have assumed even greater importance than before. Transition state theories no longer eclipse collision theories to the extent suggested by many textbooks. Both view-points have their use in the development of a full understanding of the dynamics of chemical change. New experimental techniques have been used to re-study known reactions and to investigate entirely new ones. Many anomalies in the literature data of kinetics have been resolved and more accurate detailed knowledge of the rates of a much wider range of elementary reactions has been obtained. The principles that underly the factors which determine the rates of elementary homogeneous gas reactions and the ways these may combine together in overall chemical changes seem now to be clear, although many details remain to be discovered. The rapid expansion of research in kinetics has made it increasingly difficult to provide a summary of the present state of knowledge in all detailed aspects of this field at undergraduate level that is at the same time both adequate and of a reasonable length. By restricting the field to homogeneous gas kinetics and concentrating on the principles, both theoretical and experimental, I have tried to provide a short modern text that covers all the basic material needed by undergraduates for present-day honours degree courses in gas kinetics. I have based this book on a course of lectures given at the University of Sussex to students majoring in chemistry, chemical physics, theoretical chemistry, materials science, or biochemistry in the School of Molecular Sciences. I have omitted large important areas of kinetics such as heterogeneous catalysis and solution kinetics and considered only a very few examples of gas-phase reactions with complex mechanisms. These latter were chosen to illustrate basic principles with the minimum of factual information. My aim has been to cover the fundamental principles of the subject in some depth. I have tried to discuss aspects of these principles that I feel are often incorrectly treated, glossed over, or ignored in elementary textbooks. I hope that as a result this book will prove to be useful to both students and teachers.

I have supplied answers to the problems in the belief that this will make them of more use to the reader from an instructional point of view. I have provided lists of suggested further reading of other texts and review articles at the end of each chapter. These should give the reader an adequate introduction to the very extensive literature of this subject and by a less biased and more simple route than could be achieved by detailed referencing in the present text. The chapters should be read in the order in which they are presented with the possible exception of Chapter 2 which may be read later if preferred.

G. L. PRATT

CHAPTER ONE

INTRODUCTION AND BASIC DESCRIPTIVE KINETICS

1.1 INTRODUCTION

1.1a The scope of kinetics

Chemical kinetics consists basically of the study of the rate at which changes occur in chemical systems. Kinetics therefore introduces, as it were, the variable *time* into chemistry. Much of physical chemistry is concerned with the study of the structures and other static properties of molecules into which the concept of time does not enter. The difference between these non-kinetic aspects of physical chemistry and chemical kinetics is essentially the same as the difference between statics and dynamics in the realm of mechanics. Thermodynamics for example deals only with systems in equilibrium, that is to say systems which do not change with time, consequently the variable *time* does not occur in thermodynamics. In the field of quantum mechanics many of the applications to chemistry are the result of studying stationary solutions of the time-independent wave equation. Thus the time-dependent wave equation,

$$\mathcal{H}\Psi = i\hbar \frac{\partial \Psi}{\partial t} \tag{1.1}$$

is reduced to the time-independent form,

$$\mathcal{H}\Psi' = E\Psi' \tag{1.2}$$

by the substitution of trial solutions of the form,

$$\Psi = \Psi' \exp\left(\frac{E}{i\hbar} t\right) \tag{1.3}$$

and hence predictions about molecular structures etc. are made. The limitations of these static methods can be made clear by a few simple examples. The conversion of diamond into graphite is accompanied by a decrease in free energy at normal temperatures and pressures. In the thermodynamic sense therefore this process is 'spontaneous'. Yet the rate

1

of this 'spontaneous' process is so slow that no observable change takes place over millions of years. Again consider the reaction,

$$2N_{2(g)} + 5O_{2(g)} + 2H_2O_{(l)} \rightleftharpoons 4HNO_{3(aq)} \qquad [1.1]$$

The known standard state free energies of the substances involved in this reaction allow us to calculate the equilibrium constant for this reaction thermodynamically. The result of this calculation shows that for the partial pressures of nitrogen and oxygen in the atmosphere the reaction should proceed to the right to quite an appreciable extent at normal temperatures. Thus thermodynamically we expect that the earth's atmosphere should dissolve in the sea to produce a moderately strong solution of nitric acid. Again, over periods of millions of years this reaction has not reached equilibrium. At the other extreme we have chemical reactions, such as some reactions between oppositely charged ions, which seem to go to completion almost instantaneously, for example in times of the order of 10^{-15} sec. While thermodynamics tells us the *direction* of any spontaneous change it can say nothing about the *rate* at which that change will occur. To predict this rate is the aim of chemical kinetics.

An ideal theory of chemical kinetics would start with the time-dependent wave equation which would first be solved to predict the rates of such simple physical processes as changes in the energy state of a molecule and energy transfer reactions in which no net chemical change occurs but energy is transferred between degrees of freedom or between molecules. Then elementary detailed chemical reactions in which reactants in specified quantum states are converted to chemically different products in specified quantum states would be treated similarly. By statistical averaging of the rates of these elementary detailed reactions over all possible quantum states the overall rate of an elementary chemical reaction irrespective of the detailed states of the reactants and products would be obtained. From these rates of elementary chemical steps the rates of normal complex chemical reactions could be compounded. In practice owing to the mathematical complexities of such a method a less ambitious route must be adopted. This involves two main steps. Firstly the static molecular properties, which in principle at least can be obtained by solution of the time-independent wave equation, are related to the rates of simple elementary chemical processes by *theories of reaction rates*. Secondly, the rates of these simple chemical steps are related by what is called the *mechanism* of the reaction to the observable rates of normal complex chemical processes. Similarly reversing the process experimental kinetic studies can be used firstly to deduce the reaction mechanism and secondly to test the theories of the elementary steps involved. To understand this we must first make clear the difference between elementary and complex reactions.

1.1b Elementary and complex reactions

An elementary chemical reaction is one which proceeds on a molecular scale as the result of chemically identical molecular acts, such as encounters, which convert reactant molecules into product molecules. Most normal chemical reactions that can be easily studied experimentally do not occur by the repetition of such identical single molecular processes. Many different concurrent and consecutive elementary steps must occur before the overall transformation of reactants into products as described by the stoichiometric equation is accomplished. Such reactions which are compounded of several elementary steps are called complex. The mechanism of a complex reaction is the assembly of elementary steps that go to compose it. In order to understand the kinetics of complex reactions it is necessary to analyse the overall process into its constituent elementary reactions, that is, to determine its mechanism. The kinetic consequences of this mechanism can then be interpreted in terms of the kinetic properties of the elementary reactions and of the way they are combined together in the mechanism. The kinetic properties of the elementary reactions themselves must be understood by studies of their detailed mechanics on both molecular and statistical levels.

A complication that may be mentioned briefly here is that some reactions which are chemically elementary in the sense explained above, i.e. they involve only a single chemical process, may nevertheless be resolved into more than one type of molecular act. This occurs when physical processes, such as energy transfer between two molecules, which do not involve any *chemical* changes, are an essential part of the elementary chemical process. This will be considered in more detail later (section 3.6) but for the moment we must note that these chemically elementary reactions may exhibit complex kinetics that are more characteristic of complex reactions.

1.1c Molecularity of elementary reactions

The molecularity of an elementary chemical reaction is defined as the number of molecules involved in each individual molecular act that results in the transformation of reactants into products. The term molecule is here to be interpreted as meaning any separate species involved in the reaction including for example free radicals, free atoms and ions as well as normal molecules. Clearly as so defined the term molecularity can only be used to describe *elementary* chemical reactions. Complex reactions which are composed of many different elementary reactions which may have different molecularities *do not have a molecularity*. Molecularity is also obviously restricted to being a positive integer and is a theoretical concept not an experimental number (contrast *order* section 1.3a). A reaction in which single isolated molecules of reactant, by some process of internal reorganization (e.g. of their energy) or internal motion spontaneously convert themselves into products (e.g. by isomerization or decomposition) has a molecularity of unity and is said to be *unimolecular*. These elementary steps are usually written,

$$A \rightarrow B \tag{1.2}$$

or

$$A \rightarrow B + C \tag{1.3}$$

Bimolecular reactions, which have a molecularity of two and are by far the most common type of elementary reaction, are those reactions which proceed as a result of collisions between two reactant molecules thus,

$$A + B \rightarrow products \tag{1.4}$$

Termolecular reactions have a molecularity of three and if they exist at all (see section 3.3) are comparatively rare. They involve the simultaneous collision of three reactant molecules,

$$A + B + C \rightarrow products \tag{1.5}$$

No gas-phase reactions with molecularities other than 1, 2 or 3 are known. In condensed phases, such as solutions, all reactions strictly speaking are n-molecular where n is a large integer since every molecule interacts with a large number of neighbours all the time. The equations [1.2] to [1.5] above are not simply stoichiometric equations expressing the simplest ratios of moles of reactants to products they actually represent the way the chemical processes occur on a molecular scale.

1.2   DEFINITION OF RATE OF REACTION

Since chemical kinetics is the study of the rates of chemical reactions we must define at the outset exactly what is meant by reaction rate. Clearly in a qualitative way the rate of a reaction is a measure of the rate of change with respect to time of the amounts of the substances involved in the reaction due to the fact that chemical reaction is taking place. Consider a chemical reaction for which the overall stoichiometric equation can be written as,

$$0 = \sum_{i=1}^{m} \nu_i X_i \tag{1.6}$$

where X_i are the reactants and products, ν_i are simple integral or fractional stoichiometric coefficients and m is the number of different chemical substances involved (*not* to be confused with the molecularity of an elementary reaction). For example the stoichiometric equation for the reaction between hydrogen and oxygen to form water,

$$2H_2 + O_2 = 2H_2O \qquad [1.7]$$

can be rearranged to have the form of [1.6] thus,

$$0 = -2H_2 - O_2 + 2H_2O \qquad [1.8]$$

By convention the ν_i are taken as positive for products and negative for reactants as written in the stoichiometric equation, (R.H.S. and L.H.S. respectively). The reaction rate is positive when reactants are converted to products. If we denote the number of moles (mass divided by molecular weight) of the substance X_i present in the system at a time t by n_i, and if the system is *closed* so that no transfer of matter into or out of it can occur then it follows from the stoichiometric equation [1.6] that the quantity,

$$R = \frac{1}{\nu_i}\frac{dn_i}{dt} \qquad i = 1, \ldots m \qquad (1.4)$$

is the same for all the substances involved in [1.6]. This quantity R is therefore a suitable measure of the total rate of reaction in the system. If the system is not closed then clearly allowance must be made for the net flow of X_i into or out of the system when calculating the rate of change of n_i *due to the chemical reaction.*

Had we chosen to represent the stoichiometric equation differently, say by,

$$0 = \sum_{i=1}^{m} \frac{\nu_i}{a} X_i \qquad [1.9]$$

obtained by dividing [1.6] throughout by an arbitrary number a, then the quantity representing the total rate of reaction would be from (1.4)

$$R' = \frac{a}{\nu_i}\frac{dn_i}{dt} \qquad i = 1, \ldots m \qquad (1.5)$$

It is therefore essential when defining the rate of a particular chemical process to write the stoichiometric equation for the reaction explicitly so that the arbitrary number a is specified. It is not sufficient for example simply to say that the rate of the reaction between hydrogen and oxygen is R. One must say for example that the rate of reaction [1.7] is R, and this would be exactly one half the rate of the reaction,

$$H_2 + \tfrac{1}{2}O_2 = H_2O \qquad [1.10]$$

obtained by dividing [1.7] throughout by 2. If the stoichiometry of the reaction under consideration is unknown then one particular substance, either a product or a reactant, must be chosen arbitrarily, say X_k, and the rate defined as,

$$R'' = \frac{dn_k}{dt} \tag{1.6}$$

i.e. setting

$$a = \nu_k \tag{1.7}$$

which is unknown. The substance X_k chosen must clearly be specified when making any statement about the reaction rate.

The total rate R so defined will depend on the size of the reaction system. If two identical systems were considered as a whole the total rate of reaction in the combined system would be twice that of either. It is often convenient to eliminate this extensive character of the total rate by dividing by the volume of the system V. Thus we have for the rate per unit volume,

$$r = R/V \tag{1.8}$$

$$= \frac{1}{V\nu_i} \frac{dn_i}{dt} \qquad = 1, \ldots m \tag{1.9}$$

from (1.4). This quantity r is usually called simply the *rate* of reaction for short. The dimensions of the rate, r, defined by (1.9) are (moles)(length)$^{-3}$ (time)$^{-1}$. The system of units used throughout this book will be to measure rates in moles cm^{-3} sec^{-1}. The reasons for this particular choice of units will become apparent later (pp. 113 and 148).

If in addition to being closed the system is also of constant volume then from (1.9),

$$r = \frac{1}{\nu_i} \frac{d(n_i/V)}{dt} \tag{1.10}$$

since

$$\frac{dV}{dt} = 0 \tag{1.11}$$

Hence,

$$r = \frac{1}{\nu_i} \frac{dC_i}{dt} \tag{1.12}$$

where C_i is the molar concentration of X_i, measured in moles cc^{-1}. This equation (1.12) is often taken as the definition of rate of reaction. However, many systems of interest in gas kinetics, for example flow reactors, (section 2.4b) do not satisfy (1.11) so that the more correct definition, equation

(1.9), must be used. A particularly simple case is that of a gas reaction occurring in a closed vessel of fixed volume with the gas mixture in thermal and mechanical equilibrium (no temperature or pressure gradients). Assuming the mixture is ideal then,

$$p_i V = n_i RT \qquad (1.13)$$

where p_i is the partial pressure of component X_i. Hence,

$$C_i = n_i/V$$
$$= p_i/RT \qquad (1.14)$$

Thus at constant temperature the rate of reaction is, from (1.12) and (1.14)

$$r = \frac{1}{\nu_i RT} \frac{dp_i}{dt} \qquad (1.15)$$

In much gas-kinetics work the concentrations of reactants etc. are specified in terms of their partial pressures in torr (mm of mercury) at specified temperatures since these are the quantities that are actually measured. The conversion between the two sets of units, torr (pressure) and moles cc^{-1} (concentration) is straightforward since the ideal law (1.13) applies to a sufficiently good approximation in most cases. R equals $8\cdot2054 \times 10^{-2}$ l. atmosphere deg^{-1} $mole^{-1}$ or $62,361$ cc torr deg^{-1} $mole^{-1}$.

1.3 DEPENDENCE OF RATE ON CONCENTRATION

For almost all chemical reactions that have been studied it is found that the rate of reaction depends on the concentrations of the reactants. In some cases the rate is found to depend also on the concentrations of the products and on the concentrations of substances that do not appear in the overall stoichiometric equation for the reaction (such as intermediate products, catalysts and inhibitors, chemically inert additives etc.). For complex reactions the *experimental rate law* which expresses the dependence of reaction rate on the concentrations is often a very complex function (see Chapter 4 for examples). For some simple reactions this dependence can be expressed as a simple power law for each concentration thus

$$r = k \prod_{i=1}^{s} C_i^{x_i} \qquad (1.16)$$

The product is taken over all s substances whose concentration affects the reaction rate (this number is not the same as m in [1.6] nor is it the same as the molecularity). The proportionality constant k is called the specific rate constant or coefficient, or often for short simply the *rate constant.* Occasionally it is loosely referred to as the 'rate' but this can lead to confusion and should be avoided. For these simple reactions the rate constant is independent of all the concentrations that are included in the product term (since these dependences of the rate have been factored out) but is a function of other variables such as temperature, shape of reaction vessel etc. that affect the reaction rate. Elementary reactions under certain conditions (section 1.5) follow a rate law of the form (1.16) accurately and some complex reactions follow such a law over restricted ranges of conditions to a good approximation. Provided the rate of reaction r is correctly defined as in (1.9) [not (1.12)] this rate law applies equally to both constant and variable volume systems (i.e. using concentrations $C_i = n_i/V$ on the R.H.S. of (1.16) even when $dV/dt \neq 0$).

1.3a Orders of reaction

The exponents x_i in equation (1.16) are called the *partial orders* of reaction with respect to the substances X_i. The sum of these,

$$x = \sum_{i=1}^{s} x_i \tag{1.17}$$

is called the *total* or *overall order* of reaction. Taking logarithms of (1.16) we get,

$$\log r = \log k + \sum_{i=1}^{s} x_i \log C_i \tag{1.18}$$

Thus a double logarithmic plot of rate versus the concentration C_i, all other variables being kept constant, should be a straight line if the rate law (1.16) is obeyed. The slope of this line will be the partial order x_i and the intercept on the log r axis will be $\log k + \sum_{j \neq i} x_j \log C_j$. It follows that the partial orders of reaction x_i may be defined by the alternative (and experimentally more useful) expression,

$$x_i = \left(\frac{\partial \log r}{\partial \log C_i} \right)_{C_j, T, \text{ etc.}} \tag{1.19}$$

or,

$$x_i = \frac{C_i}{r} \left(\frac{\partial r}{\partial C_i} \right)_{C_j, T, \text{ etc.}} \tag{1.20}$$

For elementary reactions under certain conditions (equilibrium energy distribution, see section 1.5) the law of mass action applies. This law states that the rate of a chemical reaction is proportional to the product of the active masses of the reacting substances. In the gas phase the active mass is proportional to the molar concentration raised to a power equal to the number of molecules of that substance involved in the elementary chemical act. The law does not apply to complex reactions (or to elementary ones in certain circumstances). Thus for an elementary unimolecular reaction [1.2] or [1.3], the rate law is,

$$r = {}_1k[A] \qquad (1.21)$$

where [A] has been written for C_A the concentration of reactant A. This is a first-order rate law and the rate constant ${}_1k$ is called the first-order rate constant. For a bimolecular reaction [1.4] the rate law is,

$$r = {}_2k[A][B] \qquad (1.22)$$

or if two identical molecules are involved,

$$A + A \rightarrow products \qquad [1.11]$$

then,

$$r = {}_2k[A]^2 \qquad (1.23)$$

These expressions (1.22) and (1.23) are second-order rate laws and ${}_2k$ is the second-order rate constant. For a termolecular reaction [1.5] the rate law is,

$$r = {}_3k[A][B][C] \qquad (1.24)$$

This is a third-order rate law and ${}_3k$ is the third-order rate constant. By comparing the rate laws (1.21) to (1.24) with that of (1.16) it is seen that for these elementary processes under suitable conditions the total order of reaction x is numerically equal to the molecularity. Because of this the terms order and molecularity were at one time used interchangeably since it was thought that the law of mass action applied to all chemical reactions. This is now known not to be so since most chemical reactions are complex and it is therefore important always to distinguish between the two quantities.

Order like molecularity is a pure number and has no units but unlike molecularity which is a theoretical concept, order is an experimental number. It simply expresses how the experimentally observed rate of reaction depends on concentration. It can therefore in principle take all values between $+\infty$ and $-\infty$. *It need not be integral.* Negative values of the order x_i imply that the substance X_i acts to slow down the rate of reaction i.e. to inhibit it. As will be seen later (section 4.3h) the action of inhibitors is usually complex so that the simple rate law (1.16) is not a very good description of the rate dependence for such systems. For other reactions it is usually found experimentally that x_i takes values between about 0 and 10, and for a great many reactions the limits are even closer, say about 0 to 3. This is to be contrasted with molecularity which can only take the integral values 1, 2 and 3.

As outlined above the majority of chemical reactions do not follow the rate law (1.16) exactly. Nevertheless this equation and those derived from it above can still be used to describe their kinetic behaviour as measured experimentally. In these cases the values x_i and k will not be truly constant but will be functions of the concentrations. Over limited concentration ranges the variation in these values will be small. Thus we may approximate the experimental plots of log (rate) versus log (concentration) by straight lines provided the curvature is not too great. The slopes and intercepts of these best straight line approximations to the experimental points give average values of the orders of reaction and the rate constant over the range of concentrations studied. Equation (1.19) can therefore be taken as the general definition of order of reaction for all reactions whether they obey the law (1.16) or not.

The units of the rate constant k can be seen from (1.16) to depend on the total order of reaction x. Measuring the rate r in moles cc^{-1} sec^{-1} and the concentrations C in moles cc^{-1} it is easily seen that the units of k are $cc^{(x-1)}$ $moles^{(1-x)}$ sec^{-1}. Only in the special case $x = 1$ are the units of the rate constant independent of the units of concentration that are used. For a complex reaction whose total order x varies with concentration it is seen that the units and hence the numerical value of the rate constant also vary with concentration. This somewhat unusual use of the term 'constant' is conventional but should not be taken too literally. In such cases it is often convenient to choose arbitrarily a convenient value for x, say for example the nearest integer, and to calculate the x-th order rate constant. This rate constant can then be tabulated as a function of concentration with the advantage that since a constant value of x has been used the *units* of this rate constant do not vary. For example, if the experimental plot of log (rate) versus log (concentration) has a slope (the order) which varies between say $1 \cdot 0$ and $1 \cdot 5$ over the range of concentration studied then a convenient way of expressing the experimental results is to calculate, say, the first-order rate constant, $_1k$, where from (1.21)

$$_1k = r/[A] \qquad\qquad (1.25)$$

This 'constant' will not in fact be constant but its variation with concentration can be tabulated and its *units* will be constant, in this case sec^{-1}. Whenever it is necessary to stress the total order of reaction x that has been taken when calculating a rate constant we shall use the convention of prefixing this to k as in the above examples, thus $_xk$.

1.3b Integration of the rate law

For a closed constant-volume system where (1.12) applies the rate law (1.16) may be written,

$$\frac{1}{\nu_i}\frac{dC_i}{dt} = k \prod_{i=1}^{s} C_i^{x_i} \qquad i = 1, \ldots m \tag{1.26}$$

This is a set of differential equations relating the concentrations C_i to time t in terms of the reaction parameters ν_i, x_i and k. Provided these parameters are constant, independent of time, these differential equations may be integrated to obtain the variations of C_i with t from known initial conditions when $t = 0$. The functions $C_i(t)$ thus obtained will depend on the orders x_i in particular. Hence the experimental data for the variation of concentrations with time in a closed constant-volume system can be used to determine the orders of reaction x_i by comparison with the integrated rate laws, $C_i(t)$. The order of reaction determined in this way is called the *order with respect to time* or for short the *time-order*. For reactions which follow the rate law (1.16) exactly throughout the whole course of the reaction with constant values of ν_i, x_i and k, clearly the time-orders must be the same as the *true orders* x_i within experimental error. The necessity for distinguishing time-order from true order arises because most chemical reactions do not satisfy these conditions exactly. Thus for example reactions with complex rate laws not of the form (1.16) will have true orders x_i and rate constants k which are functions of concentration as outlined above. These parameters therefore will vary with time as the concentration varies. Also in complex reactions as the composition of the reacting system changes and products replace reactants the whole character of the reaction mechanism may change so that the rate law itself changes completely. Even the stoichiometric coefficients ν_i may change with time in complex reactions. As a simple example of these complications consider a reaction whose rate is directly proportional to the concentrations of the single reactant A and that of a chemically inert gas M which is not consumed in the reaction, (see section 3.6)

$$A + M \rightarrow B + M \tag{1.12}$$

so that,

$$r = {}_2k[A][M] \tag{1.27}$$

The true total order of reaction which is determined by the variation of rate with concentration is seen from (1.27) to be two. When the reaction proceeds in a closed constant-volume system however only the reactant concentration [A] will change with time since M is chemically inert. Hence the time-order obtained by comparing the observed variation of reaction rate with time with the general integrated rate laws will be unity. This is a rather trivial example but it does serve to stress the importance of distinguishing true and time-orders for complex reactions. Owing to these complications the use of integrated rate laws for complex reactions is very limited and such reactions are usually studied under more closely controlled conditions than are obtained simply by allowing the reaction to proceed to large extents of reaction in closed systems.

For simple reactions these complications do not occur and true and time-order are the same. Let us consider some simple cases. Suppose that we study in a closed constant-volume system a reaction whose rate depends on the concentration of one reacting substance only. Thus, (1.26) can be written in this case,

$$-\frac{dC}{dt} = kC^x \qquad (1.28)$$

where we have taken the stoichiometric coefficient to be -1 since C is a reactant concentration. This law will hold for the special case of stoichiometric mixtures of several reactants for which the concentration *ratios* do not vary (p. 17). This rate law will also hold for reactions involving more than one reactant provided that the concentrations of these other reactants (and any other substances which affect the reaction rate) are kept constant in time so that their concentrations can be accumulated into the time-invariant rate constant k in (1.28). In practice this can be achieved in open systems by adding or removing these other substances at rates that exactly balance their rates of consumption or production by the reaction, or in closed systems approximately by making these substances be present in very large excess so that the fractional change in their concentrations due to the reaction can be neglected. The rate law (1.28) can be integrated from the assumed initial conditions,

$$C = C_0 \qquad \text{at} \qquad t = 0 \qquad (1.29)$$

as follows,

$$\int_0^t k\,dt = -\int_{C_0}^C C^{-x}\,dC \qquad (1.30)$$

Since k and x are constants we get,

$$kt = \ln(C_0/C) \qquad \text{if} \qquad x = 1 \qquad (1.31)$$

or

$$kt = (1-x)^{-1}(C_0^{1-x} - C^{1-x}) \qquad \text{if} \qquad x \neq 1 \qquad (1.32)$$

The variation of concentration with time predicted by these integrated rate laws is illustrated in Figure 1.1 for a few integral values of the order. For a zero-order reaction, (1.32) shows that the variation of C with t is linear i.e. the rate does not change with time, as shown in Figure 1.1. Clearly in practice the rate law must change as the concentration approaches zero since the line cannot cross the time axis into regions of

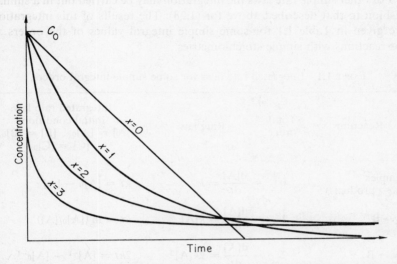

FIGURE 1.1 Plots of integrated rate laws (1.31) and (1.32) for some integral orders x. The rate constants used are; $x = 0$, $k = 1$; $x = 1$, $k = 2\cdot5$; $x = 2$, $k = 10$; $x = 3$, $k = 200$.

negative concentration. Zero-order reactions are always complex processes. For a first-order reaction (1.31) can be re-written as

$$C = C_0\, e^{-kt} \qquad (1.33)$$

so that the concentration follows a simple exponential decay with a time constant (the time taken for the concentration to be reduced to $1/e$ th of its value at any time) τ given by,

$$\tau = 1/k \qquad (1.34)$$

A plot of $\ln C$ versus t will be linear for a first-order reaction. For a second-order reaction (1.32) becomes,

$$kt = 1/C - 1/C_0 \qquad (1.35)$$

A plot of $1/C$ versus t is therefore linear. As the order of reaction increases Figure 1.1 shows that the variation of C with t becomes more and more sharply curved. However as can be seen the shape of these plots is relatively insensitive to *small* changes in x. While it may be easy to distinguish between say first and second order decay curves by their shape, to distinguish similarly between say a $1 \cdot 30$ order reaction and $1 \cdot 40$ order, one would require unusually accurate data over a wide range of concentrations.

For other simple rate laws the integration may be carried out in a similar fashion to that described above for (1.28). The results of this integration are given in Table 1.1 for some simple integral values of the orders x_i for reactions with simple stoichiometries.

TABLE 1.1   Integrated rate laws for some simple integral orders.

Reaction	Total order	Rate law	Integrated rate law, initial conditions $[A] = [A]_0$, $[B] = [B]_0$, $[C] = [C]_0$
complex ($A \rightsquigarrow$ products)	0	$-\dfrac{d[A]}{dt} = k$	$kt = [A]_0 - [A]$
$A \rightarrow B$	1	$-\dfrac{d[A]}{dt} = k[A]$	$kt = \ln ([A]_0/[A])$
$2A \rightarrow B$	2	$-\dfrac{d[A]}{dt} = 2k[A]^2$	$2kt = [A]^{-1} - [A]_0^{-1}$
$A + B \rightarrow C$	2	$-\dfrac{d[A]}{dt} = k[A][B]$	$kt = ([A]_0 - [B]_0)^{-1}$ $\ln ([A][B]_0/[A]_0[B])$
$3A \rightarrow B$	3	$-\dfrac{d[A]}{dt} = 3k[A]^3$	$3kt = \frac{1}{2}([A]^{-2} - [A]_0^{-2})$
$2A + B \rightarrow C$	3	$-\dfrac{d[A]}{dt} = 2k[A]^2[B]$	$kt = ([A]_0 - 2[B]_0)^{-1}$ $([A]_0^{-1} - [A]^{-1}) +$ $([A]_0 - 2[B]_0)^{-2} \ln$ $([A][B]_0/[A]_0[B])$
$A + B + C \rightarrow D$	3	$-\dfrac{d[A]}{dt} = k[A][B][C]$	$kt = \{([B]_0 - [C]_0) \ln$ $([A]/[A]_0) + ([C]_0 - [A]_0)$ $\ln ([B]/[B]_0) + ([A]_0 - [B]_0)$ $\ln ([C]/[C]_0)\}$ $([A]_0 - [B]_0)^{-1} ([B]_0 -$ $[C]_0)^{-1}([C]_0 - [A]_0)^{-1}$
complex ($A \rightsquigarrow$ products)	x	$-\dfrac{d[A]}{dt} = k[A]^x$	$kt = (x - 1)^{-1} ([A]^{1-x} -$ $[A]_0^{1-x})$

1.3c Measurement of order of reaction

We will consider the principles of the four main methods by which orders
of reaction are commonly measured. We will deal with these in increasing
order of unreliability.

(i) Initial rates
As outlined in section 1.3b above the variation of rate of reaction with
time in closed constant-volume systems will for simple reactions be due
solely to the depletion of the reactants by the reaction. For complex
reactions this is not always so. Interaction between the products of reaction
and the reactants or between the mechanism of subsequent reactions of the
products and the initial *mechanism* of the reaction occurs in many cases.
The variation of rate with time in such systems is very complex and is in
no simple way related to the kinetics of the initial reaction. In any system
undergoing chemical change it is only initially when the system is first
made-up from presumably pure reactants that the composition is known
accurately. Once appreciable amounts of reaction have occurred uncertainty exists about the nature of the substances present and their
concentrations. For these reasons the measurement of reaction rate is best
carried out in the first place at the initial state of the system to obtain the
initial rate. The more complex problem of what happens later on in the
reaction can be tackled after the mechanism of the initial reaction has been
determined.

To measure the initial rate entails measurement of the amount of
reaction during the very early stages of the reaction. A very sensitive analytical method is therefore required. Suitable techniques are discussed in
Chapter 2. Using such sensitive techniques the change in concentration
ΔC_i of any species involved in the reaction can be measured in a short
interval of time Δt. This time interval must be such that the extent of
reaction occurring within it is very small, say much less than 1% of the
total change for complete reaction. The initial rate,

$$r_0 = \left(\frac{1}{\nu_i}\frac{dC_i}{dt}\right)_{t=0} \tag{1.36}$$

can then be approximated by,

$$r_0 = \frac{1}{\nu_i}\frac{\Delta C_i}{\Delta t} \tag{1.37}$$

provided that $\Delta C_i/\Delta t$ does not vary significantly as Δt is decreased. If
$\Delta C_i/\Delta t$ does vary then $(dC_i/dt)_{t=0}$ can be obtained by extrapolating
$\Delta C_i/\Delta t$ to $\Delta t = 0$. An alternative but less accurate procedure is to plot

C_i versus t and to draw the initial tangent at $t = 0$. If the rate varies very rapidly during the initial stages, as occurs for example in some non-steady state reactions (see section 1.6c), this method of initial rates may not work very well. Having measured the initial rate for a given set of conditions one of the initial concentrations is changed and the new initial rate measured. In a series of such experiments the initial rate as a function of one initial concentration can be measured keeping the other initial concentrations constant. The order of reaction x_i with respect to the substance X_i whose concentration was varied is then found by plotting $\log (r_0)$ versus $\log (C_i)_0$ according to equation (1.18)

$$\log r_0 = \{\log k + \sum_{j \neq i} x_j \log (C_j)_0\} + x_i \log (C_i)_0 \qquad (1.38)$$

where $(C_i)_0$ is the *initial* concentration of substance X_i. The slope of this plot *by definition* is x_i. Similarly from the results of similar series of experiments in which the other initial concentrations are varied the other partial orders can be calculated, as also can the rate constant k. The orders of reaction thus measured will be *true* orders. If the experimental order plots are not linear then the slope at any point gives the order of reaction for that particular concentration and hence the variation of order with concentration may be studied.

In some cases it is not convenient to use this method of initial rates. An analytical technique of sufficient sensitivity may not be available or the reaction may be too fast to study at very small percentages of reaction. Other less satisfactory methods must then be used to obtain the orders and the rate constant which involve studying the reacting system after appreciable amounts of reaction have occurred. The method of fractional life-times is such a method.

(ii) Fractional life-times

In this method the time taken for a certain fixed fraction (such as 10%, 20% or 50%) of the reaction to occur is measured as a function of the initial concentrations. Assuming that the variation of rate with time is due solely to the decrease in reactant concentration the rate law may be used in its integrated forms, described in section 1.3b, to determine the orders of reaction. For example, for a reaction in which only one concentration variable occurs in the rate law, equation (1.28) applies in closed constant-volume systems. The integrated forms (1.31) and (1.32) can be rearranged to

$$t = k^{-1} \ln (C_0/C) \qquad \text{if} \quad x = 1$$

or

$$t = k^{-1}(1 - x)^{-1}\{C_0^{1-x} - C^{1-x}\} \qquad \text{if} \quad x \neq 1$$

Let the time for a *fixed* fraction, f, of reaction be t_f then these equations give,

$$t_f = k^{-1} \ln (1/1 - f) \qquad \text{if} \quad x = 1 \qquad (1.39)$$

or

$$t_f = k^{-1} C_0^{1-x} (1 - x)^{-1} \{1 - (1 - f)^{1-x}\} \qquad \text{if} \quad x \neq 1 \qquad (1.40)$$

since $C/C_0 = 1 - f$.

In either case, for fixed f, the variation of t_f with initial concentration C_0 is given by,

$$t_f \propto C_0^{1-x} \qquad (1.41)$$

or taking logarithms,

$$\log t_f = (1 - x) \log C_0 + \text{constant} \qquad (1.42)$$

Therefore a double logarithmic plot of the fractional life-time versus the initial concentration should yield a straight line of slope $(1 - x)$. Hence the order of reaction with respect to the reactant whose concentration is varied is obtained. The rate constant may be determined from the intercept provided the fixed fraction f has a known value.

Reactions with more complicated rate laws than (1.28) can often be simplified by choosing suitable conditions so that (1.28) holds for this restricted set of experiments. As mentioned previously for a stoichiometric mixture, that is one in which the reactants are present in the same proportions as those occurring in the stoichiometric equation, the ratios of reactant concentrations remain constant at these stoichiometric ratios throughout the reaction, so that any rate law *which depends only on reactant concentrations*

$$r = k \prod_{i=1}^{r} C_i^{x_i} \qquad (1.43)$$

becomes

$$r = \left\{ k \prod_{i=1}^{r} \left(\frac{\nu_i}{\nu_1} \right)^{x_i} \right\} C_1^{x} \qquad (1.44)$$

for example, where x is the total order defined by (1.17). This is of the same form as (1.28) so that the total order can be measured by studying the variation of fractional life-time with the concentration of stoichiometric mixtures. To obtain the partial orders the reaction must be studied under conditions in which all concentrations except one are kept effectively constant as described previously (p. 12).

If for some reason it is not possible to vary the initial reactant concentrations then the orders of reaction may be estimated from the way in which the rate varies with time in a closed constant-volume system. As discussed in section 1.3b orders determined in this way are time-orders, and may not be the same as the true orders. There are two methods of evaluating time-orders called the differential and integral methods.

(iii) Differential method

The measured variation with time of the concentration of reactants or products is differentiated with respect to time either (a) graphically by drawing tangents to a smooth curve fitted to the experimental points, or (b) numerically, for example by taking small finite differences $\Delta C/\Delta t$ to approximate (dC/dt) at the middle of the interval. This procedure yields the rate of reaction as a function of time and since the concentrations are related to time in a measured fashion the rate as a function of concentration is also known. For example, considering the simple case discussed above where only one concentration parameter varies the order of reaction and rate constant are obtained from the slope and intercept of a double logarithmic plot of the rate versus the concentration. In more complex cases the method is more difficult to apply. This is because all the concentrations in the reaction mixture are varying together in a dependent manner. There is only one truly independent variable during a run, that is time. Hence only one order of reaction can be measured from the time-variation. It is first necessary to guess the ratios of the orders involved if the method is to be used for a reaction with several partial orders. For example consider a reaction whose rate law involves two partial orders,

$$r = kC_A^{x_A}C_B^{x_B} \tag{1.45}$$

Taking logarithms,

$$\log r = \log k + x_A \left\{ \log C_A + \frac{x_B}{x_A} \log C_B \right\} \tag{1.46}$$

If the ratio (x_B/x_A) is known from other experiments or can be guessed then $\log r$ can be plotted against $\{\log C_A + (x_B/x_A) \log C_B\}$ and x_A and k obtained from the slope and intercept. Clearly the guesswork method is only suitable when the orders are simple integers or fractions. As before, by using stoichiometric mixtures the rate law may be simplified to (1.44) and hence the overall order of reaction measured. If the measured data on the time-variation of the concentrations are not sufficiently accurate to enable differentiation to be carried out, the integral method may be used.

(iv) Integral method

This is the worst method of all for measuring orders and should only be used if the other methods are precluded e.g. by shortage of data. This method can only be used for simple integral or fractional orders and is very inaccurate. The rate law is integrated for assumed values of the orders using the known stoichiometric equation for the reaction to provide the relations between the various concentration changes, as was done for some simple examples in section 1.3b. This integrated rate law is then compared either graphically or numerically with the experimental results to see whether a reasonable fit, using a constant value for the rate constant k, can be obtained. If a reasonable fit is found then the assumed values for the orders represent a possible set of values and the constant k is the corresponding rate constant. By trial and error the best fit for simple orders can be found. The method is very inaccurate since as pointed out above the shapes of the integrated rate curves (Figure 1.1) are relatively insensitive to small changes in order. For example, a first-order reaction should give a straight line when ln (concentration) is plotted against time. If the data for a reaction whose correct order is, say, 1·5 are plotted out in this form quite a reasonable straight line results if only a limited concentration range is covered, say 0–50% reaction. Using this method then it would be concluded that the reaction was first order. Many examples of such an erroneous identification of reaction order by this method exist in the chemical literature. The only advantage of the integral method is that it requires very few (and not particularly accurate) experimental results. It is of very little use for most gas-phase reactions but is more frequently used in solution kinetics.

1.4 DEPENDENCE OF RATE ON TEMPERATURE

Many reactions are studied under conditions of thermal equilibrium (see section 2.2a) in which the distribution of molecules over their available energy levels is described by the Boltzmann distribution law, equation (A.21), in terms of a single variable T, the absolute temperature. The overall observed rate of an elementary chemical reaction is an average over all the detailed energy states of the reactants, each of which will have very different rates of reaction. As the temperature of the system is varied the distribution of reactants over the energy levels varies rapidly [see (A.21)] and so the observed average rate constant also varies markedly. Experimentally it was found by Hood in 1878 that the variation of reaction rate with temperature could be obtained in a linear form by plotting the logarithm of the rate against the reciprocal of the absolute temperature. This plot has since been found to be linear for a very large number of reactions. It is therefore convenient to express the variation of the rate constant with temperature by means of the *experimental activation energy* E_a which is defined by,

$$E_a = - \frac{d \ln k}{d(1/RT)} \qquad (1.47)$$

where R is the gas constant, and E_a is usually expressed in units of kcal mole^{-1}. E_a is also sometimes called the Arrhenius activation energy or simply the activation energy. The justification for these names for what is essentially just the temperature coefficient of the reaction rate will appear later (Chapter 3). For the present we will simply note that E_a has the dimensions of energy per mole. For many reactions, notably elementary ones, the value of E_a defined in this way is very nearly constant, independent of temperature within experimental error. Normally E_a lies in the range from about -10 kcal mole^{-1} to $+100$ kcal mole^{-1} for most chemical reactions. Negative values of E_a imply that the reaction rate decreases as the temperature increases and are comparatively rare (section 3.3). The majority of chemical reactions have positive values for E_a i.e. their rates increase rapidly with increasing temperature. Sometimes the plot of ln k versus $(1/RT)$ is curved so that the slope, E_a, varies with temperature. This is usually a sign that the reaction is complex, unless it so happens that E_a is very small (say a few kcal mole^{-1}) when small variations in E_a become proportionately more important.

Equation (1.47) can be integrated between $(1/T) = 0$ and $(1/T)$ to give,

$$\ln k(T) - \ln k(\infty) = - \int_0^{1/RT} E_a d(1/RT)$$

or,

$$k(T) = k(\infty) \exp\left\{ - \int_0^{1/RT} E_a d(1/RT) \right\} \tag{1.48}$$

where $k(T)$ is the value of k at temperature T. If E_a is independent of temperature this becomes

$$k(T) = k(\infty) \exp(-E_a/RT) \tag{1.49}$$

Using the symbol A in place of $k(\infty)$ this is,

$$k(T) = A \exp(-E_a/RT) \tag{1.50}$$

This is called the *Arrhenius equation*. A is called the *pre-exponential factor* and is independent of temperature (provided E_a is, as assumed above). Sometimes A is called the frequency factor but in general this name should be avoided. The units of A are seen from (1.50) to be the same as those of the rate constant k since the exponential term has no units. In the case of first-order reactions only, A has the dimensions of a frequency, e.g. sec^{-1}, but for reactions of other orders the units of concentration are involved.

Plots of ln k versus $1/T$ are usually called Arrhenius plots. If these plots are not exactly linear so that E_a varies with temperature it is conventional to define a pre-exponential factor $A(T)$, which is a function of temperature, so that the Arrhenius law is still formally obeyed. Thus, we define $A(T)$ by,

$$A(T) = k(T) \exp\left\{-\frac{1}{RT}\frac{\text{d} \ln k(T)}{\text{d}(1/RT)}\right\} \qquad (1.51)$$

From (1.47),

$$A(T) = k(T) \exp(E_a(T)/RT) \qquad (1.52)$$

so that,

$$k(T) = A(T) \exp(-E_a(T)/RT) \qquad (1.53)$$

which is the same as the Arrhenius equation (1.50) but $A(T)$ and $E_a(T)$ are now functions of temperature rather than constants. For most reactions any such variations are relatively small over restricted temperature ranges. The standard method of obtaining E_a and A from experimental data is therefore the same in all cases; measured values of $\log_{10} k$ are plotted against $1/T$ and the slope of the best straight line over the range of temperature is calculated; this is $-E_a/2\cdot303R$ where R equals $1\cdot987$ cal mole^{-1} deg^{-1}; the value of E_a obtained is then used together with the experimental values of the rate constant k to calculate A from equation (1.52).

The variation of rate constant with temperature for reactions which follow the Arrhenius law (1.50) is illustrated in Figure 1.2. k rises in a sigmoid fashion with increasing T to a limiting value of A at very high

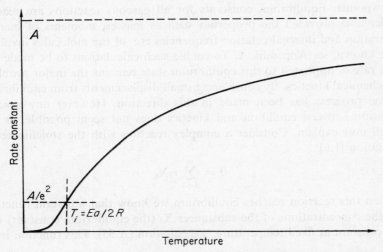

FIGURE 1.2 The variation of the rate constant with temperature for a reaction which follows the Arrhenius law (1.50).

temperatures. The temperature at the point of inflexion T_i is obtained by differentiating equation (1.50) twice with respect to temperature and equating the result to zero. The result is,

$$T_i = E_a/2R \tag{1.54}$$

Most reactions of normal stable molecules have activation energies in the range 10–100 kcal mole^{-1}. Since R is about 2 cal mole^{-1} deg^{-1}, values of T_i for these reactions lie between 2,500 and 25,000°K. For these reactions therefore most experimental studies are restricted to the region of temperature below T_i where the reaction rate increases more or less exponentially with rising temperature and the rate constant is very much smaller than its high temperature limiting value A. Reactions of very reactive species such as free radicals and atoms often have activation energies between 0 and 10 kcal mole^{-1} so that for these reactions the rate constant may approach its limiting high temperature value at experimentally accessible temperatures.

1.5 REVERSIBLE REACTIONS

1.5a Kinetics and equilibrium position

The relationship between the equilibrium state of a system and its rate of approach to that state is a problem that has received a great deal of attention. This is because the factors governing the equilibrium state of such simple systems as chemically reacting mixtures of ideal gases are now clearly understood both from thermodynamic and statistical mechanical viewpoints. Equilibrium constants for all gaseous reactions are readily calculable provided the properties such as masses, moments of inertia, vibration and internal rotation frequencies etc. of the molecules involved are known, see Appendix A. To enable such calculations to be made for the rate of approach to this equilibrium state remains the major problem of chemical kinetics. By considering small displacements from equilibrium some progress has been made in this direction. However any *general* relation between equilibria and kinetics does not seem possible, as we shall now explain. Consider a complex reaction with the stoichiometric equation [1.6]

$$0 = \sum_{i=1}^{m} \nu_i X_i$$

When this reaction reaches equilibrium we know that a certain function of the concentrations of the substances X_i (the equilibrium constant) will be constant at fixed temperature, see equation (A.55). This function is,

$$\prod_{i=1}^{m} [X_i]_e^{\nu_i} = K_c \tag{1.55}$$

where $[X_i]_e$ represents the concentration of X_i at equilibrium and K_c represents the *numerical value* of the equilibrium constant at a given temperature. Hence from (1.55) it is also true that,

$$F\left(\prod_{i=1}^{m} [X_i]_e^{\nu_i}\right) = F(K_c) \qquad (1.56)$$

$$= \text{a constant} \qquad (1.57)$$

where $F(K_c)$ is any *arbitrary* function of K_c, and is therefore also a constant. Now consider the rate at which this reaction approaches this equilibrium state from an infinitesimal distance away from equilibrium to the left. The kinetics of this reaction from left to right will be governed by a rate law which we may write quite generally as,

$$r_f = R_f([X_i]_e) \qquad (1.58)$$

where the suffixes f stand for the *forward* reaction. In simple cases (1.58) will have the form (1.16) for example. Similarly for the approach to equilibrium from a position infinitesimally away to the right we may write,

$$r_b = R_b([X_i]_e) \qquad (1.59)$$

where b stands for backwards reaction. When the system reaches equilibrium no further net change occurs so that

$$r_f = r_b \qquad (1.60)$$

Hence,

$$R_f([X_i]_e) = R_b([X_i]_e) \qquad (1.61)$$

or

$$\frac{R_f([X_i]_e)}{R_b([X_i]_e)} = 1 \qquad (1.62)$$

This is another relation between the concentrations at equilibrium and it must therefore be consistent with the thermodynamic result (1.56) hence,

$$\frac{R_f([X_i]_e)}{R_b([X_i]_e)} = \frac{F(K_c)}{F\left(\prod_{i=1}^{m} [X_i]_e^{\nu_i}\right)} \qquad (1.63)$$

or,

$$R_b([X_i]_e) = F\left(\prod_{i=1}^{m} [X_i]_e^{\nu_i}\right) R_f([X_i]_e)/F(K_c) \qquad (1.64)$$

This equation *appears* to relate the kinetic rate laws for the forward and backward reactions through the equilibrium constant. However the function F is completely arbitrary so that in fact this relation tells us

24 GAS KINETICS

nothing. This can be made clearer by a simple example. Consider the reaction,

$$A + B \rightleftharpoons C + D \tag{1.13}$$

which may be a kinetically complex reaction but has the *stoichiometry* given by [1.13]. Then the equilibrium constant is,

$$\frac{[C]_e[D]_e}{[A]_e[B]_e} = K_c \tag{1.65}$$

Suppose, for example, that the forward reaction has been found experimentally to follow the simple rate law,

$$r_f = k_f[A][B] \tag{1.66}$$

Then very close to equilibrium,

$$r_f = k_f[A]_e[B]_e \tag{1.67}$$

At equilibrium,

$$r_f = r_b \text{ as before,}$$

therefore

$$r_b = k_f[A]_e[B]_e \tag{1.68}$$

From (1.65), i.e. taking $F(K_c) = K_c$, and (1.68),

$$r_b = \left(\frac{k_f}{K_c}\right)[C]_e[D]_e \tag{1.69}$$

Assuming that this kinetic rate law does not vary as we move away from equilibrium we can write non-equilibrium concentrations in (1.69) thus,

$$r_b = \left(\frac{k_f}{K_c}\right)[C][D] \tag{1.70}$$

We *appear* to have derived the kinetic rate law for the backward reaction from the experimental law for the forward rate. However, instead of taking the equilibrium constant as in (1.65) we could equally well for example take,

$$\frac{[C]_e^2[D]_e^2}{[A]_e^2[B]_e^2} = K_c^2 \tag{1.65'}$$

$$= K_c' \tag{1.71}$$

i.e. taking $F(K_c) = K_c^2$ which is still a constant. In this case (1.68) and (1.72) gives,

$$r_b = \left(\frac{k_f}{K_c'}\right) \frac{[C]_e^2 [D]_e^2}{[A]_e [B]_e} \tag{1.69'}$$

or far from equilibrium, assuming the same law holds,

$$r_b = \left(\frac{k_f}{K_c'}\right) [C]^2 [D]^2 [A]^{-1} [B]^{-1} \tag{1.72}$$

This is a completely different rate law from (1.70). Since $F(K_c)$ can be chosen arbitrarily the form of the rate law 'deduced' is arbitrary also. Hence in general for complex reactions the existence of the equilibrium constant established themodynamically does not imply any relation between the kinetics of the forward and reverse reactions near to equilibrium let alone under conditions far removed from equilibrium. If however the *mechanism of the reaction at equilibrium is known* then the rate laws for the forward and backward reactions together with the condition of equilibrium (1.60) yield a relation between the rate constants and the equilibrium constant for conditions close to equilibrium. We can illustrate this by considering the important special case of elementary reactions.

1.5b Elementary reactions

The fundamental mechanical laws which govern the motions of particles on a microscopic scale are invariant under *time reversal*. That is to say replacing the time t by $-t$ does not change the equations of motion. This is true irrespective of whether the motion is governed by quantum or classical mechanics. Application of this fact to a molecular act such as a collision between two molecules leads to the *principle of microscopic reversibility*. This states that if in the final state after such a collision all the momenta (including those of any internal motions of the molecules as well as their translational motion) are reversed then the system will *exactly* retrace its path in the reverse direction to form a state similar to the original initial state but with reversed momenta. In other words on a microscopic scale all processes can go in either direction without violating the laws of motion. This is of course not true for macroscopic systems which consist of statistically large numbers of microscopic systems. A cine film of everyday life when run backwards does not show a feasible series of events but on a microscopic scale it would. Application of this principle to statistically large assemblies of molecules *in equilibrium* leads to a result known as the *principle of detailed balancing* of elementary reactions. This states that at equilibrium every molecular process proceeds *on average* equally fast in both directions. This means that cyclic equilibria are impossible on a molecular scale. For example if three substances A, B, and C are in equilibrium, the equilibrium cannot be maintained by a cycle such as,

$$A \longrightarrow B$$

[1.15]

$$C$$

but each step must be balanced separately thus,

$$A \rightleftarrows B$$

[1.14]

$$C$$

Thus it follows that for an elementary reaction the path taken must be the same both forwards and backwards. The law of mass action for elementary reactions also follows from the application of the microscopically reversible laws of motion to statistically large assemblies of molecules in equilibrium. Thus considering for example the *elementary* (contrast [1.13]), reversible reaction,

$$A + B \rightleftarrows C + D \tag{1.16}$$

the law of mass action gives,

$$r_f = k_f[A][B] \tag{1.66}$$

$$r_b = k_b[C][D] \tag{1.73}$$

By the principle of detailed balancing at equilibrium,

$$r_f = r_b \tag{1.60}$$

therefore

$$k_f[A]_e[B]_e = k_b[C]_e[D]_e$$

or

$$\frac{[C]_e[D]_e}{[A]_e[B]_e} = \frac{k_f}{k_b} \tag{1.74}$$

From the definition of the equilibrium constant K_c, (1.55), this equation is,

$$K_c = \frac{k_f}{k_b} \tag{1.75}$$

This equation is sometimes referred to as the principle of detailed balancing since it is derived directly from this principle as stated above for elementary reactions in equilibrium.

However no system undergoing net chemical reaction can be exactly in equilibrium. The occurrence of reaction must alter the distribution of molecules over their available energy states. It has now been shown that provided the translational energy distribution of the molecules is not significantly disturbed from the equilibrium Maxwellian distribution, then detailed balancing and the law of mass action apply to elementary reactions. This is true whether or not disturbance from equilibrium for the other types of energy, e.g. vibrational energy, occurs. For most systems of chemical interest the disturbance to the Maxwell velocity distribution is negligible unless it is deliberately altered (e.g. section 2.5c), so that detailed balancing and the law of mass action *do* apply to elementary reactions in these systems to a sufficiently good approximation. Equation (1.75) may therefore be applied to systems of *known mechanism* even though the chemical reaction may disturb the equilibrium population of molecules over their internal energy states as is a fairly common occurrence (e.g. section 3.6).

1.5c The Arrhenius equation

The result of detailed balancing, equation (1.75), may be used to derive some relations between thermodynamic quantities and the parameters of the Arrhenius equations for reversible reactions of known mechanism. Taking natural logarithms of (1.75), gives

$$\ln k_f - \ln k_b = \ln K_c \tag{1.76}$$

The thermodynamic equation for K_c is,

$$\Delta G_c^\circ = -RT \ln K_c \tag{1.77}$$

where ΔG_c°, the standard free energy change for the reaction, is the change in Gibbs free energy when reactants in their standard states are converted to products in their standard states in accordance with the stoichiometric equation as written. The suffix c denotes that the standard states are defined in terms of unit *concentration*. It is more usual for gases to define the standard state in terms of unit pressure (1 atmosphere). The corresponding equation to (1.77) is,

$$\Delta G_p^\circ = -RT \ln K_p \tag{1.78}$$

where K_p is the equilibrium constant in terms of pressures. K_p is related to K_c by using the equation (1.14) for ideal gases and substituting into equation (1.55), thus,

$$K_c = \prod_{i=1}^{m} [X_i]_e^{\nu_i}$$

$$= \prod_{i=1}^{m} p_{i,e}^{\nu_i} (RT)^{-\sum_{i=1}^{m} \nu_i} \tag{1.79}$$

$$= K_p (RT)^{-\Delta \nu} \tag{1.80}$$

where $\Delta v = \sum\limits_{i=1}^{m} v_i$ is the change in number of molecules in the stoichiometric equation.

Equations (1.76) and (1.77) give,

$$\ln k_f - \ln k_b = -\Delta G_c^\circ / RT \tag{1.81}$$

This standard free energy change may be written in terms of the separate enthalpy and entropy contributions,

$$\Delta G_c^\circ = \Delta H_c^\circ - T\Delta S_c^\circ \tag{1.82}$$

so that,

$$\ln k_f - \ln k_b = \frac{-\Delta H_c^\circ}{RT} + \frac{\Delta S_c^\circ}{R} \tag{1.83}$$

The Arrhenius equations (1.50) for the forward and backward rate constants can be written,

$$k_f = A_f e^{-E_{af}/RT}$$
$$k_b = A_b e^{-E_{ab}/RT} \tag{1.84}$$

Substituting these expressions into (1.83) gives,

$$-\frac{(E_{af} - E_{ab})}{RT} + \ln (A_f/A_b) = \frac{-\Delta H_c^\circ}{RT} + \frac{\Delta S_c^\circ}{R} \tag{1.85}$$

Broadly speaking, the first term on the L.H.S. correlates with the enthalpy term on the right and the second term on the L.H.S. correlates with the entropy term. The correlation is not however an exact equality. The correct relation may be derived as follows. The van't Hoff isochore may be written,

$$\frac{d \ln K_p}{d(1/RT)} = -\Delta H_p^\circ \tag{1.86}$$

Equation (1.80) for K_c gives,

$$\frac{d \ln K_c}{d(1/RT)} = \frac{d \ln K_p}{d(1/RT)} + \Delta v RT \tag{1.87}$$

Hence we get

$$\frac{d \ln K_c}{d(1/RT)} = -(\Delta H_p^\circ - \Delta v RT) \tag{1.88}$$

$$= -\Delta E_p^\circ \tag{1.89}$$

since for ideal gases,

$$\Delta(PV) = \Delta v RT$$

For ideal gases the standard state symbols can be omitted from the enthalpy change in (1.86) and the energy change in (1.89) since the enthalpy and energy depend on temperature only and not on pressure. Now by definition the activation energies for the forward and backward reactions are, [equation (1.47)],

$$E_{af} = - \frac{d \ln k_f}{d(1/RT)}$$

$$E_{ab} = - \frac{d \ln k_b}{d(1/RT)}$$

(1.90)

Differentiation of (1.76) with respect to $(1/RT)$ and substitution of (1.90) and (1.89) gives,

$$E_{af} - E_{ab} = \Delta E \tag{1.91}$$

or using (1.88) in place of (1.89) gives,

$$E_{af} - E_{ab} = \Delta H - \Delta v RT \tag{1.92}$$

In practice for many gas reactions at normal temperatures $\Delta v RT$ is small compared to ΔH.

The corresponding relationship between the pre-exponential factors and the entropy change is then given by substituting equation (1.92) into (1.85),

$$\ln (A_f/A_b) = \frac{\Delta S_e^\circ}{R} - \Delta v \tag{1.93}$$

For standard states defined in terms of pressures we can derive the corresponding equation as follows.
Equations (1.76) and (1.80) give,

$$\ln k_f - \ln k_b = \ln K_p - \Delta v \ln (RT) \tag{1.94}$$

Substitution for K_p from equation (1.78) gives,

$$\ln k_f - \ln k_b = \frac{-\Delta G_p^\circ}{RT} - \Delta v \ln (RT) \tag{1.95}$$

Again writing,

$$\Delta G_p^\circ = \Delta H_p^\circ - T\Delta S_p^\circ \tag{1.96}$$

we get,

$$\ln k_f - \ln k_b = \frac{-\Delta H_p^\circ}{RT} + \frac{\Delta S_p^\circ}{R} - \Delta v \ln (RT) \tag{1.97}$$

Using equations (1.84) for the rate constants,

$$-\frac{(E_{af} - E_{ab})}{RT} + \ln{(A_f/A_b)} = \frac{-\Delta H_p^\circ}{RT} + \frac{\Delta S_p^\circ}{R} - \Delta \nu \ln{(RT)} \quad (1.98)$$

and substituting (1.92) into this as before, we get,

$$\ln{(A_f/A_b)} = \frac{\Delta S_p^\circ}{R} - \Delta\nu(1 + \ln{(RT)}) \quad (1.99)$$

The equations (1.93) and (1.99) which relate pre-exponential factors to entropy changes suggest an alternative way of formulating the experimental Arrhenius equation, which is sometimes convenient. Thus we may write,

$$A_f = \mathscr{F} \; e^{n_f} \; e^{S_{af}/R} \quad (1.100)$$

$$A_b = \mathscr{F} \; e^{n_f} \; e^{S_{ab}/R} \quad (1.101)$$

where n_f is the total number of reactant molecules in the stoichiometric equation and n_b is similarly the total number of product molecules, i.e.

$$n_f = - \sum_{\text{reactants}} \nu_i \quad (1.102)$$

$$n_b = + \sum_{\text{products}} \nu_i \quad (1.103)$$

and $\mathscr{F}$ is a constant for all reactions—a kind of 'universal pre-exponential factor'. S_{af} and S_{ab} are called the 'entropy of activation' for the forward and backward reaction respectively by analogy with the term energy of activation for E_a, since substitution of (1.100) and (1.101) into (1.93) gives the result,

$$S_{af} - S_{ab} = \Delta S_c^\circ \quad (1.104)$$

which may be compared with (1.91). Using this symbolism the Arrhenius equation becomes,

$$k = \mathscr{F} \; e^n \; e^{S_a/R} \; e^{-E_a/RT} \quad (1.105)$$

Assuming $\mathscr{F}$ is independent of temperature, taking logarithms and differentiating with respect to $1/RT$ gives,

$$\frac{d \ln k}{d(1/RT)} = \frac{1}{R}\frac{dS_a}{d(1/RT)} - E_a - \frac{1}{RT}\frac{dE_a}{d(1/RT)} \quad (1.106)$$

$$\therefore \qquad T\frac{dS_a}{d(1/RT)} = \frac{dE_a}{d(1/RT)} \quad (1.107)$$

or

$$T\frac{dS_a}{dT} = \frac{dE_a}{dT} \quad (1.108)$$

Thus the variation (if any) of S_a and E_a with temperature are related in the same way as are those of the thermodynamic quantities ΔS_c° and ΔE, that is,

$$T\frac{d\Delta S_c^\circ}{dT} = \frac{d\Delta E}{dT} = \Delta C_v \qquad (1.109)$$

As will be seen later in connexion with transition state theories of elementary reactions (section 3.13) this nomenclature for S_a may in certain cases gain more than a purely formal significance, but in general this is not the case. For complex reactions of unknown mechanism S_a and E_a may still be defined as above in terms of the experimentally observed reaction rates but they are not in any way related to a ΔS_c° and ΔE for any process that may be clearly specified. For these reasons it is more usual to express the experimental results in terms of A and E_a.

It is sometimes argued that equation (1.83) can be used to *derive* the Arrhenius equation by splitting this equation into two parts one involving properties of the reactants and one involving properties of the products. This argument however is fallacious. The equation may be split into two in an infinite number of ways. The Arrhenius equation is only one of an infinite set of possible equations that are consistent with the thermodynamic equations for the equilibrium constant. Justification for the Arrhenius equation can only come from kinetic evidence, either as an experimental fact or as a result that is predicted by numerous theories of elementary reactions.

1.5d Small displacements from equilibrium

Consider, as an example, the reversible elementary reaction [1.16], initially at equilibrium. If we make a very small displacement Δz from the equilibrium position towards the right by means of some small disturbance of the conditions then,

$$[A] = [A]_e - \Delta z$$
$$[B] = [B]_e - \Delta z \qquad (1.110)$$
$$[C] = [C]_e + \Delta z$$
$$[D] = [D]_e + \Delta z$$

The rates of the forward and backward reactions are given by (1.66) and (1.73) so that the net forward rate of the reaction is,

$$r = r_f - r_b \qquad (1.111)$$
$$= k_f[A][B] - k_b[C][D] \qquad (1.112)$$

and from (1.110),

$$r = \frac{-d[A]}{dt} = \frac{-d[B]}{dt} = \frac{d[C]}{dt} = \frac{d[D]}{dt} = \frac{d\Delta z}{dt} \qquad (1.113)$$

Substitution of (1.113) and (1.110) into (1.112) gives,

$$\frac{d\Delta z}{dt} = k_f([A]_e - \Delta z)([B]_e - \Delta z) - k_b([C]_e + \Delta z)([D]_e + \Delta z) \qquad (1.114)$$

Expanding these products and using equation (1.74) we get,

$$\frac{d\Delta z}{dt} = -\Delta z\{k_f([A]_e + [B]_e) + k_b([C]_e + [D]_e)\} + O(\Delta z^2) \qquad (1.115)$$

For vanishingly small displacements we can ignore the terms of order Δz^2 and write,

$$\frac{d\Delta z}{dt} = -k\Delta z \qquad (1.116)$$

where,

$$k = k_f([A]_e + [B]_e) + k_b([C]_e + [D]_e) \qquad (1.117)$$

Equation (1.116) represents a first-order rate equation for Δz which may be integrated as in section 1.3b to give the typical exponential decay law for a first-order process,

$$\frac{\Delta z}{\Delta z_0} = e^{-kt} \qquad (1.118)$$

where Δz_0 is the initial displacement at $t = 0$. The time constant for this decay, $1/k$, is called the relaxation time since it governs the rate at which the system returns to equilibrium or *relaxes* after the initial small displacement. Measurement of this relaxation time experimentally enables both k_f and k_b to be calculated, from equations (1.117) and (1.75).

Although we have discussed only one particular case of a reversible elementary reaction [1.16] the above treatment can clearly easily be extended to any equilibrium reaction provided the mechanism is known. Whatever the detailed rate laws for the forward and backward reactions are, they may be expanded as power series in terms of the small displacement about the equilibrium position as was done in (1.115) above. For very small displacements all terms other than the first order can be neglected to a first order of approximation. The zero-order term is always zero by virtue of the equilibrium condition. Hence the return to equilibrium always follows a first-order decay law (1.118) to a first order of approximation irrespective of the detailed kinetic laws of the forward and backward reactions.

1.6 COMPLEX REACTIONS

We have seen in the preceding sections that elementary reactions have many relatively simple kinetic properties. They obey the law of mass action so that their order is numerically equal to their molecularity, they follow the Arrhenius law to a good approximation, and they obey the principle of detailed balancing so that forward and reverse processes are related by the thermodynamics of the reaction. Complex reactions in general do not have these simplifying features. We now wish to see how the kinetics of complex reactions may be deduced from those of the elementary reactions of which they are built-up. That is we seek the relation between the kinetics of a complex reaction and its mechanism.

Complex reactions may consist of any number of elementary steps that may be opposing reactions (reversible), concurrent reactions (in parallel) or consecutive reactions (in series) compounded together in any way. The rate laws for each step can be written very simply by employing the law of mass action and hence it is a relatively simple matter to write down the differential rate laws governing the rates of change of concentration of all the substances involved in the reaction whether they be reactants, final products or intermediate products, of whatever reactivity. The integration of these differential equations in order to obtain for example the time variation of these concentrations in a closed constant-volume system presents in general a formidable problem in mathematics. Even for elementary reactions the integrated rate laws are often not very simple expressions, see Table 1.1. For complex reactions consisting of just a pair of opposing or concurrent or consecutive reactions, the equations can be integrated straightforwardly to obtain in general rather cumbersome integrated rate expressions. We shall not go into the details of these integrations as they illustrate no new principles. Examples of these calculations can be found in the references suggested for further reading at the end of this chapter. We shall consider in detail only one very simple case, that of two consecutive first-order reactions, and we shall use this to introduce an approximate method for dealing with these complex mechanisms that introduces very great simplification and is of very wide applicability.

1.6a Two consecutive first-order reactions

The mechanism can be represented as,

$$A \xrightarrow{\ k_1\ } B \xrightarrow{\ k_2\ } C \qquad\qquad [1.17]$$

where A is the reactant, B an intermediate product and C the final product. The rate constants for the two steps are k_1 and k_2 as represented in [1.17]. The intermediate B is formed from A and undergoes further reaction itself to produce C so that the *net* rate of formation of B is clearly the difference between its rate of formation by step 1 and its rate of decay by step 2. Thus for a closed constant-volume system we can write down the following differential equations for the rates of change of the concentrations of the three substances A, B, and C,

$$-\frac{d[A]}{dt} = k_1[A] \tag{1.119}$$

$$\frac{d[B]}{dt} = k_1[A] - k_2[B] \tag{1.120}$$

$$\frac{d[C]}{dt} = k_2[B] \tag{1.121}$$

since both steps are assumed to be first-order processes. If we take the initial conditions as,

$$[A] = [A]_0$$
$$[B] = [C] = 0 \tag{1.122}$$
$$t = 0$$

the integration of these simultaneous differential equations is (in this case) quite simple. Integration of (1.119) gives for [A],

$$[A] = [A]_0 \, e^{-k_1 t} \tag{1.123}$$

which is a simple first-order decay law. Substituting this result into (1.120) gives a differential equation involving the concentration of B only i.e.

$$\frac{d[B]}{dt} = k_1[A]_0 \, e^{-k_1 t} - k_2[B] \tag{1.124}$$

This linear first-order inhomogeneous differential equation integrates to,

$$[B] = [A]_0 \left(\frac{k_1}{k_2 - k_1} \right) (e^{-k_1 t} - e^{-k_2 t}) \tag{1.125}$$

as can easily be verified by differentiating and substituting back into (1.124). The variation of [C] with time can then finally be obtained either by substituting (1.125) into (1.121) to give,

$$\frac{d[C]}{dt} = k_2[A]_0 \left(\frac{k_1}{k_2 - k_1} \right) (e^{-k_1 t} - e^{-k_2 t}) \tag{1.126}$$

and integrating this, or perhaps more simply, by using the conservation of mass equation,

$$[A]_0 = [A] + [B] + [C] \qquad (1.127)$$

which shows that,

$$[C] = ([A]_0 - [A]) - [B]$$

which from (1.125) and (1.123) gives,

$$[C] = [A]_0 (1 - e^{-k_1 t}) - [A]_0 \left(\frac{k_1}{k_2 - k_1}\right)(e^{-k_1 t} - e^{-k_2 t}) \qquad (1.128)$$

The variations of the concentrations with time given by (1.123), (1.125) and (1.128) are plotted out in Figure 1.3 for various relative magnitudes of the two rate constants k_1 and k_2. To clarify the shapes of these curves they have been plotted on different concentration scales for the three substances in the cases where $k_2 \gg k_1$. As can be seen from this figure when $k_2 \gg k_1$ the intermediate B reaches an almost steady concentration after a time of the order of $(1/k_2)$. This steady concentration is very small being given by $[B]/[A]_0 \simeq (k_1/k_2)$. The rate of formation of product C therefore also becomes approximately steady at $k_1[A]$ after an *induction period* $(1/k_2)$. These conclusions can be made clear by dividing (1.125) by (1.123) to get the ratio of the concentrations of B and A, thus,

$$[B]/[A] = \left(\frac{k_1}{k_2 - k_1}\right)(1 - e^{-(k_2 - k_1)t}) \qquad (1.129)$$

For $k_2 \gg k_1$ this becomes to a good approximation,

$$[B]/[A] = \frac{k_1}{k_2}(1 - e^{-k_2 t}) \qquad (1.130)$$

This ratio therefore rises to a steady value k_1/k_2 with a time constant of $(1/k_2)$.

If we were unable to obtain the exact solutions of (1.119) to (1.121) for a closed constant-volume system as above we could simplify the problem by considering an open system in which we add reactant A at a rate equal to its rate of reaction. In this way we could keep the concentration of A in the system constant. For this experimental system the rate equations become,

$$\frac{d[A]}{dt} = r - k_1[A] = 0 \qquad (1.131)$$

$$\frac{d[B]}{dt} = k_1[A] - k_2[B] \qquad (1.120)$$

$$\frac{d[C]}{dt} = k_2[B] \qquad (1.121)$$

where r is the rate at which A is added to the system per unit volume.

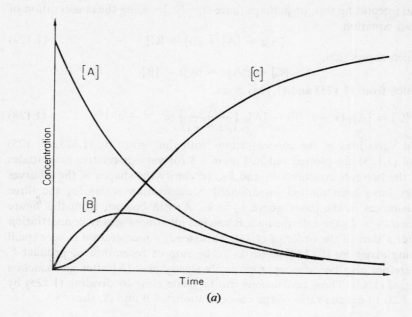

(a)

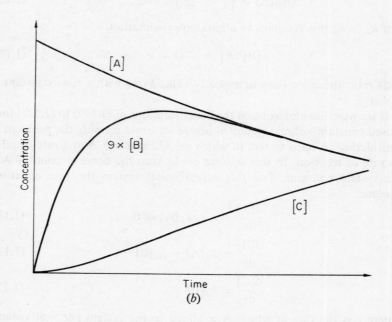

(b)

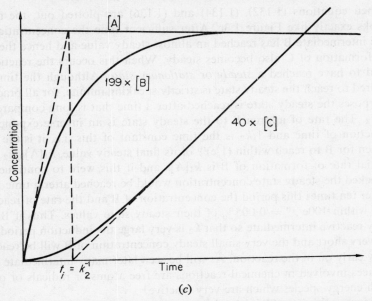

$t_i = k_2^{-1}$

Time

(c)

FIGURE 1.3   Plot of concentrations versus time for the reaction with mechanism
$A \xrightarrow{k_1} B \xrightarrow{k_2} C$ in a closed constant volume.

(a) $k_2 = 2k_1$
(b) $k_2 = 10k_1$. [B] has been multiplied by $(k_2 - k_1)/k_1 = 9$ for clarity.
(c) $k_2 = 200k_1$. [B] has been multiplied by $(k_2 - k_1)/k_1 = 199$, and [C] by 40 for clarity.

From (1.131),

$$[A] = [A]_0 \qquad \text{at all } t \qquad\qquad (1.132)$$

Substituting into (1.120)

$$\frac{d[B]}{dt} = k_1[A]_0 - k_2[B] \qquad\qquad (1.133)$$

so that,

$$[B] = \frac{k_1}{k_2}[A]_0(1 - e^{-k_2 t}) \qquad\qquad (1.134)$$

Substituting this result into (1.121),

$$\frac{d[C]}{dt} = k_1[A]_0(1 - e^{-k_2 t}) \qquad\qquad (1.135)$$

which is readily integrated to give,

$$[C] = k_1[A]_0 t + \frac{k_1}{k_2}[A]_0(e^{-k_2 t} - 1) \qquad\qquad (1.136)$$

When equations (1.132), (1.134) and (1.136) are plotted out, the result looks exactly like Figure 1.3c. After sufficient time the concentration of the intermediate B has reached an almost steady value and hence the rate of formation of C also becomes steady. When this occurs the reaction is said to have reached a *steady* or *stationary state*. Although the time required to reach the steady state is strictly speaking infinite, for all practical purposes the steady state is reached after a time that is long compared to $1/k_2$. The rate of approach to the steady state is an inverse exponential function of time and $1/k_2$ is the time constant of this, i.e. it is the time taken for B to reach within $(1/e)^{\text{th}}$ of its final steady value, $k_1[A]_0/k_2$. The initial rate of formation of B is $k_1[A]_0$ and if this were to continue unchecked the steady state concentration would be reached after a time $1/k_2$. After ten times this period the concentration of B and the rate of reaction are within $100e^{-10} = 0.005\%$ of their steady state values. Thus if B is a very reactive intermediate so that k_2 is very large the induction period will be very short and the very small steady concentration of B will be reached very early on in the reaction. As will be seen later many intermediate substances involved in chemical reactions are free atoms or radicals or other high energy species which are very reactive.

For conditions sufficiently close to the steady state, equations (1.131), (1.120) and (1.121) become to a very good approximation,

$$\frac{d[A]}{dt} = r - k_1[A] = 0 \qquad (1.137)$$

$$\frac{d[B]}{dt} = k_1[A]_0 - k_2[B] = 0 \qquad (1.138)$$

$$\frac{d[C]}{dt} = k_2[B] = k_1[A]_0 \qquad (1.139)$$

Thus by setting the rate of change of the concentration of intermediate equal to zero we have reduced the problem of solving the simultaneous differential equations to one involving only the solution of simultaneous *algebraic* equations which is a much easier process. In this case we obtain the solution that the rate of reaction as measured by formation of the final product C is given in the steady state by (1.139). Thus if we repeat the experiment in the open system with different values for the concentration of A we shall find that the rate of reaction is first order with respect to A with a rate constant equal to that for the initial step, k_1. Alternatively we could allow the concentration of A to vary very slowly. Provided this rate of variation is very slow compared to the rate of approach to the steady state then the stationary concentration of the intermediate will continually adjust itself to the appropriate value given by (1.138) as the concentration of A changes. Returning now to experiments in the closed constant-volume system where the concentration of A is varied by virtue of its depletion by reaction with a time constant $1/k_1$ we can see that provided $1/k_1 \gg 1/k_2$ i.e. $k_2 \gg k_1$ the reaction proceeds for most of the time, i.e. after an induction period of the order of $1/k_2$, only negligibly far displaced from the steady state. In this simple case this conclusion is easily verified by reference to the exact solutions (1.123) to (1.130) of the differential equations (1.119) to (1.121) for a closed system.

1.6b The steady state hypothesis or method

In the simple example given above application of the steady state method is unnecessary and trivial but in more complex situations where several intermediate products are involved it is often the only method of solving the problem of predicting the kinetics of a complex reaction from its mechanism. The method used to apply the steady state hypothesis is in general as follows. The differential rate laws are written down for all the substances involved in the reaction. The rates of change of concentration of all the very reactive intermediates are set equal to zero. This yields a number of simultaneous algebraic equations for the concentrations of these intermediates. Since the number of equations is always the same as the number of intermediate concentrations the solution of these equations to obtain the intermediate concentrations in terms of the concentrations of reactants and final products is always possible although the resulting formulae may be rather complicated in complex cases. Having obtained the steady state concentrations of the intermediates the rate law for the reaction can then be formulated in terms of the concentrations of the reactants and stable products only.

For the steady state approximation to apply it is necessary that the intermediates which are assumed to reach steady states should be very reactive species such as atoms and free radicals or highly excited molecules. The rate constants for their destruction must greatly exceed those for their formation so that their concentrations must be very low compared to those of the other species present. The low concentration of these intermediates is often used as a starting point for arguments leading to the steady state solution. Thus it is sometimes stated that since the concentration of B in the above example is small, then $d[B]/dt$ is also small and hence can be set approximately equal to zero. However this argument, as it stands, is misleading. The rate of reaction in the above example is equal to $k_2[B]$ and because the rate constant k_2 is very large the concentration of B is small in systems where the rate of reaction is not large. Nevertheless if the concentration of B were to be doubled, say, then the rate of reaction also would double no matter how small the absolute value of [B]. That is, the time variation of B, $d[B]/dt$, is of fundamental importance in determining how the rate of reaction varies. The only equation in which $d[B]/dt$ can be neglected simply because [B] is so small is that representing the conservation of mass, i.e. in the above example,

$$\frac{d[A]}{dt} + \frac{d[B]}{dt} + \frac{d[C]}{dt} = 0 \tag{1.140}$$

where all the other terms are very much greater. The ability to set $d[B]/dt$ equal to zero for the steady state solution rests on the condition that the concentration of reactant, A, is held constant or varied only very slowly.

The exact solution in the above example, equations (1.123), (1.125), (1.128), shows that only at one instant of time during the closed constant-volume reaction is it exactly correct to set $d[B]/dt$ equal to zero. This is at the maximum value of [B] which occurs at a time,

$$t_{max} = \frac{1}{(k_2 - k_1)} \ln (k_2/k_1) \tag{1.141}$$

and the maximum concentration is,

$$[B]_{max} = [A]_0 \left(\frac{k_1}{k_2}\right)^{k_2/(k_2 - k_1)} \tag{1.142}$$

as can be easily verified by differentiating (1.125) and equating to zero. But this is *not* the true steady state, which is only reached exactly after an infinite time, except in the trivial case when k_1 equals zero. Moreover this time t_{max} is not the induction period as can be seen from (1.141) if k_1 is very small. The steady state is reached to a very good approximation appreciably earlier than t_{max}. For example, the reaction is within 1% of the steady state after a time equal to

$$t_{0.99} = 4 \cdot 6 k_2^{-1} \tag{1.143}$$

but if say $k_2/k_1 = 10^8$, a not unreasonable value for reactions involving free radicals (see Chapter 4), then t_{max} is four times larger than $t_{0.99}$ since, $t_{max} = 18 \cdot 5 k_2^{-1}$.

Specific examples of the application of the steady state method will be given in Chapter 4 but for the present its use will be illustrated by one fairly simple but important example. Consider a complex reaction which proceeds via the following elementary steps,

$$A + B \xrightarrow{k_1} C + B \qquad\qquad [1.18]$$

$$C + B \xrightarrow{k_2} A + B \qquad\qquad [1.19]$$

$$C \xrightarrow{k_3} D \qquad\qquad [1.20]$$

that is, a reversible bimolecular reaction followed by a unimolecular isomerization or decomposition. Provided that C is a reactive intermediate in the sense explained above, in the steady state we may write,

$$\frac{d[C]}{dt} = 0 = k_1[A][B] - k_2[B][C] - k_3[C] \qquad (1.144)$$

since C is formed in step (1) and destroyed by both steps (2) and (3). Solving equation (1.144) for the steady state concentration of C,

$$[C] = \frac{k_1[A][B]}{k_2[B] + k_3} \qquad (1.145)$$

Thus the steady rate of reaction as measured by the rate of formation of the product D is,

$$\frac{d[D]}{dt} = k_3[C] \qquad (1.146)$$

$$= \frac{k_1 k_3[A][B]}{k_2[B] + k_3} \qquad (1.147)$$

The rate law is thus obtained in terms of the concentrations of the reactants only and does not involve the concentration of the reactive intermediate. However it does, of course, involve the rate constants for *all* the elementary steps. This rate law is not of the form of equation (1.16) since it does not involve simply the products of various powers of the reactant concentrations. The orders of reaction for this complex process are therefore not constant but vary with concentration. In certain limiting cases however it does approximate to a simple rate law of the form (1.16). If the rate of step (2) greatly exceeds that of step (3) i.e.

$$r_2 \gg r_3 \qquad (1.148)$$

so that

$$k_2[B] \gg k_3 \qquad (1.149)$$

either because k_2 is very large or else because we make the concentration
of B in the reaction mixture very large, then the rate law (1.147) becomes
approximately,

$$\frac{d[D]}{dt} = \frac{k_1 k_3 [A][B]}{k_2[B]}$$

$$= \frac{k_1 k_3}{k_2} [A] \tag{1.150}$$

This has the form (1.16) and the reaction is first order with respect to A
and zero order with respect to B. Note however that the effective first-
order rate constant is a function of all three elementary steps,

$$_1 k = \frac{k_1 k_3}{k_2} \tag{1.151}$$

Under these conditions reaction (3) is said to be the *rate-determining step*
in the overall process since it is the slowest reaction and its first-order
kinetics govern those of the overall process. If on the other hand reaction
(3) is much faster then reaction (2),

$$k_2[B] \ll k_3 \tag{1.152}$$

either because k_3 is very large or because we make the concentration of B
very small, the rate law (1.147) becomes approximately,

$$\frac{d[D]}{dt} = k_1[A][B] \tag{1.153}$$

This again has the simple form (1.16) but it is now second-order overall
being first-order with respect to both A and B. In this case reaction (1)
is the 'bottleneck' in the overall process and is said to be rate-determining.
Since it is also the first step the rate of reaction is actually *equal* to the rate
of this rate-determining step. In general for a system of consecutive reactions if one step is much slower than all *subsequent* steps it is said to be
rate-determining. The overall rate of reaction may depend on the rates of
all steps preceding the rate-determining step but will not depend on any
of the fast subsequent steps. For cases intermediate between (1.152) and
(1.149) neither step is rate-determining and the kinetics do not conform
to the simple law (1.16). The order of reaction varies with the concentration of B between one at large [B] and two at small [B]. This particular
mechanism will be considered in more detail in section 3.6 when unimolecular reactions are discussed.

1.6c Non-steady states

As will be seen in Chapter 4 when some complex reactions are considered in detail the stationary state method is of very great value in simplifying the kinetic treatment of many complex reactions. However, many other reactions are known to occur to which the stationary state hypothesis cannot be applied. This is either because the steady state is not reached during the time for which the system is observed and during which appreciable amounts of reaction occur or else because no stationary state as such exists. This latter occurrence may be due either to inherent chemical features of the mechanism (section 4.4b) or to the physical characteristics of the system, e.g. the heat balance (section 4.4a). All three of these types of non-steady state are considered in more detail in Chapter 4. For these reactions the kinetics must be evaluated by direct consideration of the simultaneous differential equations which describe the system of elementary steps involved. If necessary simplifying assumptions other than the steady state hypothesis must be made to enable a mathematical solution to be obtained. One of the most useful of these assumptions is that the reactive intermediates reach steady concentrations *relative* to one another, i.e.

$$\frac{d\{[R_1]/[R_2]\}}{dt} = 0 \qquad (1.154)$$

where R_1 and R_2 are any two reactive intermediates, but that the total concentration of reactive intermediates varies in a non-steady way. This assumption reduces the simultaneous differential equations to a series of simultaneous algebraic equations for the relative concentrations of the intermediates together with a single differential equation describing the variation in the total concentration of intermediates. Some examples of this approach will be given in Chapter 4. Bodenstein who first introduced this treatment called it the method of *quasi-stationary states*.

When many intermediates of different reactivity are involved in a reaction the sequence of events during the initial stages of the reaction is as follows. The most reactive intermediate increases in concentration very rapidly until it approaches a quasi-stationary state relative to the second most reactive intermediate. The concentrations of these two then rise more slowly in unison until they reach a quasi-stationary state with respect to the third most reactive intermediate and so on. This process continues until all the intermediates reach a quasi-stationary state which is determined by the slow rise of the most unreactive intermediate and in which all the concentrations rise in proportion together towards the final overall steady state. The kinetics during these successive quasi-stationary states

44   GAS KINETICS

are clearly very complex but they can sometimes be evaluated by numerical solution of the differential equations using either digital or analogue computors.

SUGGESTIONS FOR FURTHER READING

Text books

S. W. Benson, *Foundations of Chemical Kinetics*, McGraw-Hill, New York, 1960.

A. A. Frost and R. G. Pearson, *Kinetics and Mechanism*, Wiley, New York, 1961.

A. Weissberger, *Techniques of Organic Chemistry*, Vol. VIII, Part I, p. 107 p. 343, Wiley, New York, 1961.

Reviews

Quart. Rev., **18**, 227 (1964).
Progr. Reaction Kinetics, **2**, 337 (1964).

CHAPTER 1 PROBLEMS

1. Which of the following reactions *might feasibly* occur as elementary steps? (Whether they *actually* do so is irrelevent here). Give their molecularities.

    (a) $N_2O = N_2 + O$
    (b) $C_2H_4 + HCl = C_2H_5Cl$
    (c) $C_5H_{12} + 8O_2 = 5CO_2 + 6H_2O$
    (d) $H_2 + F_2 = 2HF$
    (e) $H + H + H_2 = 2H_2$
    (f) $Pb(C_2H_5)_4 = Pb + 4C_2H_5$
    (g) $2H + 2O = H_2O_2$
    (h) $Na + C_2H_5Cl = NaCl + C_2H_5$

Write down expressions for the rate of reaction per unit volume in a closed constant-volume system for each of the above reactions in terms of the rates of change of the concentrations of all the substances involved.

2. The initial rate of a reaction between A and B was found to vary with the pressure of A when that of B was kept constant at 10 torr as follows,

Pressure of A (torr)	10	15	25	40	60	100
$10^3 \times$ rate (torr sec^{-1})	1·0	1·22	1·59	2·00	2·45	3·16

When the pressure of A was kept constant at 10 torr and that of B varied the initial rates were,

Pressure of B (torr)	10	15	25	40	60	100
$10^3 \times$ rate (torr sec^{-1})	1·0	1·84	3·95	8·00	14·7	31·6

Find the partial orders of reaction with respect to A and B. Assuming that no other substances affect the reaction rate what is the total order? Calculate the rate constant using pressure units (torr). If the temperature is 400°C what is the rate constant using concentration units (mole cc^{-1})?

3. The initial rate of a reaction varies with the pressure of reactant as follows,

Pressure (torr)	1·0	1·41	1·78	2·24	3·35	6·31	10·0
$10^3 \times$ rate (torr sec^{-1})	1·0	2·0	3·16	5·01	11·2	23·4	42·7

Pressure (torr)	20·0	31·6	63·1	100·0
$10^3 \times$ rate (torr sec^{-1})	100·0	166	316	501

Evaluate the order of reaction with respect to this reactant at 1·0 and at 100·0 torr pressure.

4. The rate of a certain reaction depends on the concentrations of the two reactants A and B only. A standard mixture in which the ratio of A:B is constant was used to investigate the variation of initial rate with total pressure of the mixture. The results were

Total pressure (torr)	100	200	300	400	500
Rate $\times 10^3$ (torr sec^{-1})	1·26	3·00	4·94	7·12	9·42

What is the total order of reaction? Is it possible to calculate the rate constant from these data?

5. The half-life (time for 50% reaction) of a reaction between A and B varies with the concentration of A when B is present in excess as follows,

Half-life (sec)	126	110	101	95·4	88·0	83·1	79·4
Pressure of A (torr)	5	10	15	20	30	40	50

What is the order of reaction with respect to A? When mixtures of A and B in stoichiometric proportions are used the half-life varies as follows,

Half-life (sec)	251	144	104	82·9	59·9	47·6	39·8
Total pressure (torr)	50	100	150	200	300	400	500

What is the order of reaction with respect to B?

6. A stoichiometric mixture of A and B undergoes reaction in a closed constant volume. The concentration of A varies with time as follows,

$[A] \times 10^6$ (mole cc^{-1})	10	9	8	7	6	5	4	3
Time (sec)	0·0	17·1	39·7	70·8	115·2	182·8	295·2	508·4

$[A] \times 10^6$ (mole cc^{-1})	2	1
Time (sec)	1018·5	3062

Use the differential method to evaluate the overall order of reaction (time-order) graphically.

7. The decomposition of a pure substance A in a closed constant-volume system proceeds as follows.

$[A] \times 10^6$ (mole cc^{-1})	7·36	3·68	2·45	1·84	1·47	1·23
Time (sec)	0·0	60	120	180	240	300

Use the integral method to obtain the order of reaction (time-order). (You may assume that in this example the order is an exact integer.) Calculate the corresponding rate constant.

8. The thermal decomposition of nitryl chloride obeys the stoichiometric equation

$$2NO_2Cl = 2NO_2 + Cl_2$$

The reaction can be followed by measuring the pressure rise in a closed vessel or by measuring the amount of light absorbed by the NO_2 produced. Some results for the variation with time of the concentration of reactant during one run at 180°C are,

Time (t) (sec)	0	150	300	450	600	750	900
$10^7 \times [NO_2Cl]$ (mole cc^{-1})	1·99	1·80	1·63	1·47	1·33	1·21	1·09

Show by plotting a suitable graph that the variation of reactant concentration with time follows a first-order decay and calculate the first-order rate constant for the reaction (stating units). In further experiments at 180°C the initial rate of decomposition of NO_2Cl was measured for a series of initial pressures of reactant and the results were;

Initial conc^n. $10^7 \times [NO_2Cl]$ (mole cc^{-1})	0·5	1·0	1·5	2·0	2·5	3·0	3·5	4·0
Initial rate $10^{12} \times \left(\dfrac{d[NO_2Cl]}{dt}\right)_{t=0}$ (mole cc^{-1} sec^{-1})	8·5	33·5	76·5	135	211	303	413	540

Determine the true order of reaction with respect to nitryl chloride and calculate the corresponding rate constant for this temperature (stating units). Measurements of the initial rates of decomposition of a fixed concentration of 10^{-7} mole cc^{-1} of NO_2Cl at four temperatures gave the following results;

Temperature °C	180	203	227	248
Initial rate $10^{12} \times \left(\dfrac{d[NO_2Cl]}{dt}\right)_{t=0}$ (mole cc^{-1} sec^{-1})	33·5	130	623	1680

Calculate the activation energy and hence deduce the pre-exponential factor of the rate constant derived in the second part above.

9. The rate of the elementary reaction,

$$HI + C_2H_4 \rightarrow C_2H_5I$$

is $1·76 \times 10^{-2}$ torr sec^{-1} when $[HI] = [C_2H_4] = 500$ torr at 300°C. At 400°C with $[HI] = 50$ torr, $[C_2H_4] = 100$ torr the rate is $1·40 \times 10^{-2}$ torr sec^{-1}. Calculate the activation energy of the reaction. Calculate also the pre-exponential factor A of the second-order rate constant using units of ccs, moles, and seconds.

10. The half-life for a reaction between fixed concentrations of reactants varies with temperature as follows,

Temperature °C	525	540	555	570	585	600
Half life (sec)	1072	631	380	229	144·5	89·1

Evaluate the activation energy of this reaction.

11. The rate of a certain reaction was studied over a very wide temperature range and the activation energy was found to vary with temperature as follows,

Temperature °C	500	1000	1500	2000	2500
E_a kcal mole^{-1}	48·0	46·0	44·0	42·1	40·1

Show that these results are consistent with a temperature dependence of the pre-exponential factor of the form $A \propto T^n$ and determine n.

12. The first-order rate constant for a certain reaction varies with the concentration of the single reactant as follows,

Pressure of reactant (torr)	10	20	40	70	100
$10^3 \times$ first-order rate constant (sec^{-1})	1·00	1·23	1·51	1·78	2·00

Find the true order of reaction with respect to the reactant.

13. The elementary reaction

$$I + H_2 \rightarrow HI + H$$

has a second-order rate constant of $10^{14.1} \exp(-33,500/RT)$ cc mole^{-1} sec^{-1}. Given that $\Delta H° = 32·84$ kcal mole^{-1} and $\Delta S° = 2·36$ cal °K^{-1} mole^{-1} calculate the Arrhenius equation for the rate constant of the reverse reaction.

14. For the dissociation of iodine

$$I_2 \rightarrow 2I$$

at 2000°C, $\Delta H° = 38·30$ kcal mole^{-1}. Assuming that the combination of iodine atoms has zero activation energy what is the activation energy for the dissociation at 2000°C?

15. A reaction proceeds by the following mechanism,

$$A + B \underset{2}{\overset{1}{\rightleftharpoons}} C$$

$$C \overset{3}{\longrightarrow} B + D$$

Where C is a very reactive intermediate. Apply the steady state method to show that the reaction is second order overall and obtain an expression for the second-order rate constant in terms of the rate constants of the elementary steps (1)–(3). Given that these rate constants are; $k_1 = 10^{11} \exp(-10,000/RT)$ cc mole^{-1} sec^1, $k_2 = 10^{15} \exp(-40,000/RT)$ sec^{-1}, $k_3 = 10^{13} \exp(-30,000/RT)$ sec^{-1} plot an Arrhenius graph of the overall second-order rate constant for temperatures between 800 and 1400°K.

16. A reaction has the mechanism,

$$A + B \underset{2}{\overset{1}{\rightleftharpoons}} C + D$$

$$C + B \overset{3}{\longrightarrow} E + D$$

$$2D \overset{4}{\longrightarrow} F$$

Where A and B are the reactants, E and F are the final products and C and D are reactive intermediates. Assume that $r_2 \gg r_3$ and use the steady state method to obtain an expression for the rate of reaction in terms of [A] and [B]. What are the partial orders with respect to A and B? What is the total order?

CHAPTER TWO

EXPERIMENTAL METHODS

2.1 COMPLEX AND ELEMENTARY REACTIONS

Experimental investigations of the kinetics of chemical reactions may be broadly classified into two types. Firstly, there are those studies of the overall complex reactions which constitute the major part of 'normal everyday chemistry'. These reactions have complicated kinetic features and one aim of the studies is to relate these characteristics to the detailed mechanism of the reaction in terms of the elementary processes involved. In this way they may be understood in terms of the properties of these simpler steps and the patterns in which they are combined together. In such studies the initial object of the experiments is to measure as many as possible of the experimental features such as the dependence of the reaction rate, measured for all species present, reactants, intermediate products as well as final products, on concentration of every one of these species, temperature, the presence of chemically inert gases, catalysts, inhibitors, shape, size and material of the reaction vessel, method of activation, e.g. whether by heating or by irradiation, etc., and time or the extent of progress of the reaction. In this way as many criteria as possible are obtained for testing various hypotheses concerning the mechanism of the reaction. Even with all this data available it is rare for it to be possible to make an unequivocal choice of mechanism. Studies in other, related but simpler, systems must often be used to assist the choice. In particular direct studies of elementary reactions are of very great assistance.

The second broad type of kinetic study, therefore, involves the use of experiments specifically designed to study the individual elementary reactions themselves. These studies enable the relations between their kinetics and the detailed molecular dynamics of the process to be investigated and in this way lead to a better understanding of the theory of these processes. In addition these studies make available a body of quantitative rate data for elementary steps which, as stated above, are needed to help interpret complex reactions. Many of the elementary reactions that occur in the gas phase are those of very reactive species such as free atoms and radicals, ions and highly excited molecules. Such reactions are often therefore extremely rapid in comparison with the overall reactions of normal stable species. The experimental techniques used in these studies therefore often need apparatus with very fast time resolution or else some way of reducing the rate to more measurable proportions must be found. Before describing some of these techniques we will first consider, in a very general way, the methods by which a chemical reaction can be caused to occur.

2.2 METHODS OF ACTIVATION

Very few chemical reactions, other than combination reactions of atoms, free radicals or oppositely charged ions can occur at the absolute zero of temperature. In general before chemical reaction can take place some energy requirements must be satisfied. For example the molecules that are to react may need a certain minimum amount of energy and possibly this must also be distributed in a certain specific way within the molecules. There may also be other requirements, for example, that the phases of motion in the various internal degrees of freedom shall be within certain limits and so on. A necessary condition for most reactions is that the reactant molecules must possess at least a certain minimum total energy before reaction can occur. This condition is not sufficient, of course, since the total energy may not be correctly distributed over the various degrees of freedom and there may be other requirements that are not related to energy, but it is necessary. The process by which this necessary energy is acquired is called, loosely, *activation*. As will be seen later (section 3.8) in certain contexts the term *activation* has a more specific meaning than this and then the process we are describing here is more correctly called *energization*. However common useage is such that in this chapter we will use the term activation to cover any process by which the reactants acquire the energy they need to react. Clearly the reaction rate will depend on the extent of this activation, i.e. how much energy in excess of the minimum the reactant molecules have. It may also depend on the method of activation, i.e. the method by which this energy is acquired since the details of this process will affect the energy distribution and also any non-energetic factors that there may be. The methods of activation in common use may be divided into four main types all of which produce rather different types of activation. These are called thermal, photolytic, radiolytic, and chemical activation and are considered in more detail below.

2.2a Thermal activation

This is the most conventional way of causing a chemical reaction to occur and it is also in some respects the simplest. In thermal activation the molecules acquire the energy needed for activation by non-reactive collision either with other reactant molecules or inert, heat bath, molecules or with the walls of the container. This process of energy transfer by collision produces a Maxwell–Boltzmann energy distribution in the equilibrium system which is not undergoing any chemical reaction. Provided the chemical reaction is sufficiently slow it will not appreciably disturb this equilibrium energy distribution, i.e. the thermal steady state differs only insignificantly from thermal equilibrium. Under these conditions the normal statistical treatment of systems containing large numbers of molecules shows that the distribution of number of molecules (n_i) over the various available energy levels (w_i levels of energy ε_i) is described by Boltzmann's equation, [see (A.21)]

$$n_i \left/ \sum_{i=0}^{\infty} n_i = w_i \, e^{-\frac{\varepsilon_i}{kT}} \right/ \sum_{i=0}^{\infty} w_i \, e^{-\frac{\varepsilon_i}{kT}} \tag{2.1}$$

in terms of a single parameter T, the temperature of the equilibrium system. Under these conditions the degree of activation of the molecules is accurately known and readily calculated from (2.1). The overall rate of reaction is determined by the appropriate average of the rates of reaction of the individual states. Many ordinary thermal reactions are studied under these conditions and as will be seen later (Chapter 3) the theory of such reactions is greatly simplified by this lack of disturbance to the equilibrium energy distribution. This also makes for simpler experimental results. The system may be characterized by a single value of the temperature and the rate of reaction as a function of temperature determined in a straightforward fashion.

However many thermal reactions can take place in such a manner that the disturbance of the Boltzmann distribution is not negligible. This may occur either because the processes tending to establish this equilibrium distribution, i.e. collisional energy transfers, become too slow, for example as may be achieved by reducing the rate of collision by reducing the gas pressure or by making a very rapid temperature change (e.g. see 2.5a or d), or else because the chemical reaction becomes too fast. In some cases the translational energy distribution may be non-Maxwellian, as occurs for example in velocity selected molecular beams (see 2.5c) or in the very fast reactions of highly reactive ions and free atoms. Other chemical reactions do not normally disturb the translational energy distribution since translation to translation energy transfer occurs on virtually every gas-kinetic collision whereas most chemical reactions occur more slowly than this (see section 3.2). Rotational and vibrational energy distributions are more easily affected, since these are established by slower energy transfer processes. Such dis-equilibrium may take the form of an energy distribution which is Boltzmann in form but having an effective temperature that is different from the temperature which describes the translational energy distribution. For example, if vibrational to vibrational energy transfer is fast compared to the chemical reaction but translational to vibrational energy transfer is too slow to maintain equilibrium, the vibrational temperature will differ from the translational temperature. In more extreme cases the energy distributions may become non-Boltzmann in form and hence cannot be described by a temperature at all. Under these conditions, clearly, the reaction rate will be a very complex function not only of the rates of reaction of all the individual energy states but also of the rates at which these states are produced by the particular method of activation that is effective. While the equilibrium systems are simpler to study, by virtue of their averaged nature they can give less detailed information concerning the mechanics of the reaction processes involved than can the more complex non-equilibrium systems. Both types of system must be studied experimentally, using suitable apparatus and techniques, to get the full picture of any reaction. For complex reactions, in view of the need for simplification, thermal activation under conditions of thermal equilibrium is most commonly employed in initial studies.

'Mechanical' activation in which the reactant molecules are given a large directed translational velocity, for example by impact with a fast moving solid surface, may be included under the heading of thermal activation particularly since as noted above this directed momentum is generally degraded by collisions to thermal motion prior to reaction.

2.2b Photolytic activation

In photolytic activation, or photolysis, the molecules acquire the energy needed for activation by absorption of light. Einstein's law of the photochemical equivalent states that only one quantum of light is absorbed by each molecule that is involved in the primary photo-chemical reaction. This is a consequence of the low density of photons in normal light beams. In experiments with abnormally intense light sources such as pulsed lasers, multiphoton processes may be observed with very low probabilities. With normal light sources the probability of two photons striking the same molecule during the time that is available before the primary reaction occurs is negligibly small. Hence the quantum yield of the initial step, where quantum yield is defined as the number of molecules reacting per quantum of light absorbed, is unity under normal conditions of photolysis. Any deviation of the experimentally observed values for the overall quantum yield of the chemical reaction from unity therefore indicates that a complex mechanism is in operation in which the primary photo-chemical process is followed by secondary reactions. For example as will be seen in Chapter 4 the overall quantum yields of complex chain reactions give a measure of what is called the chain length of the reaction.

The primary photochemical reaction is clearly unimolecular since it involves only one molecule of reactant, plus one photon. The photon may of course only be absorbed if the energy of the quantum is equal to the difference in energy between two states of the molecule. For visible and U.V. light, transitions will be from the ground to excited electronic states. These transitions will be vertical in the Franck–Condon sense, that is the positions of the nuclei of the atoms in the molecule will not change appreciably during the absorption process. This is because the time taken for the transition is very short, typically about 10^{-16} sec, compared to the time required for appreciable nuclear movement, typical nuclear vibration periods being around 10^{-14} sec. The natural life-time of the excited electronic state is typically about 10^{-8} sec after which it will, if nothing else has happened to it in the meantime, return to the ground state by re-emitting a light quantum. For the absorption of non-monochromatic radiation the probabilities of transition to the various possible upper states will be determined by the overlap of the wave functions of the lower and upper states. The initial energy distribution of the excited molecule will therefore be very different from a Boltzmann distribution. If this initial energy distribution of each degree of freedom can be expressed by a meaningful temperature at all, such temperatures will be greatly in excess of the ambient temperature in most cases. Such excited species are often called 'hot' molecules. They may be electronically, vibrationally, or rotationally 'hot' and if fragmentation of the molecule occurs the fragments may be translationally 'hot'. This excess energy may be re-radiated (after about 10^{-8} sec), reshuffled by internal convertions or degraded by collisions with other molecules to thermal motion of the system as a whole. The period between these quenching collisions of the excited molecule with other molecules is inversely proportional to the gas pressure; at atmospheric pressure it is typically about 10^{-10} sec. Whether chemical reaction occurs before thermalization or not depends on the relative rates of the two processes. Secondary reactions in photochemical systems may in different cases therefore be either ordinary thermal reactions or may involve non-equilibrated 'hot' species. As pointed out in section 2.2a, translational and rotational energy is more rapidly exchanged than is vibrational energy so that reactions of translationally or rotationally hot molecules are less common in photochemical reactions than those of vibrationally hot species.

In many simple cases spectroscopic evidence is available to indicate what the products of the primary photochemical reaction are. Thus for example, species in specific excited electronic states may be produced by photolytic activation and their reactions studied and compared to those of the ground state. If the initial products *are* known from such evidence and if the excess energy is rapidly degraded to thermal motion then the method of photolytic activation provides a technique for studying the normal thermal reactions of such reactive species as free atoms and free radicals. However in complex systems the initial states or even in some cases the initial products may be unknown and the presence of excess energy in the initial products may lead to unwanted complications.

Many molecules do not absorb light in the conveniently accessible region of the spectrum. In these cases photosensitization may be used to activate the reactant. A species, the sensitizer, is added to the system. This sensitizer can absorb the incident radiation and can then transfer the energy by collision to the reactant molecules. The action of light on the sensitizer may be either simply excitation to a higher electronic state or it may involve subsequent dissociation of the sensitizer into reactive fragments. In the first case the electronically excited state is either quenched by collision with reactant molecules which are thereby excited or it undergoes chemical reaction with the reactant. In the second case the fragments of the primary decomposition react chemically with the reactant molecules thereby initiating reaction. The possibilities for complications are clearly increased by the presence of a sensitizer and in some cases the whole character of the reaction may be altered.

One advantage of photochemical activation as a technique is that the energy given to the reacting molecule is precisely known and can be varied by varying the frequency of the incident light. Practical difficulties may arise, however, in doing this since intense monochromatic sources, such as mercury discharge lamps, provide only a relatively few different frequencies while continuous sources used with monochromators provide only low intensities. In addition the activation process may be started and stopped with great rapidity thus enabling very fast processes to be studied. The approach of the reaction to a steady state can be investigated by using intermittent or pulsed illumination (see sections 2.5e and f). Disadvantages of the technique are the often unknown nature of the distribution of the energy in the activated reactant and the rather high values of the energy that are involved, for example a quantum of mercury resonance radiation of 2536·5Å wavelength is equivalent to an energy of 112 kcal mole^{-1} which is far more than is sufficient to initiate most normal chemical reactions.

2.2c Radiolytic activation

Molecules of reactant may be activated by collision with fast-moving particles such as electrons, protons, neutrons, alpha particles, 'hot' recoil atoms (emitted in radioactive decays), and γ-ray quanta or other forms of 'ionizing radiation'. A distinction may be drawn between these processes and normal photolysis described above in that photochemical processes usually involve transitions only to low lying electronic states whereas the greater energies available in radiolytic activation make transitions to states above the ionization level important. Thus radiolysis generally leads to ion formation in the primary step and so for convenience we may include photo-ionization under this heading. The distinction is in some sense arbitrary but is nevertheless convenient from a practical point of view. For the same reasons that were explained above for photolytic activation (only more so) the initial products of radiolytic activation, which will be ions, electrons, excited molecules and fragments, will have distinctly non-Boltzmann energy distributions. The actual distribution will be determined by the energy distribution of the incident particles and the (vertical) transition probabilities to the various possible energy states. Often very large energies, on a chemical scale, are involved and the resulting secondary reactions are far removed from the normal field of chemical reactions under thermal conditions. The interactions of ions with electrons and neutral molecules constitute a branch of gas-phase kinetics that is now much studied both practically and theoretically. In some cases (see section 3.4) these very fast ion-molecule reactions show some simple characteristics which can be interpreted by relatively simple theories of their detailed mechanics.

The ions formed initially undergo further reactions of great rapidity so that equilibration of the internal energy over the various energy levels and between the various degrees of freedom is unlikely to occur before chemical reaction takes place. In very low pressure systems such as mass spectrometers (section 2.3b) some success has been achieved in predicting theoretically the rates of secondary unimolecular reactions of ions by using the assumption that internal energy equilibration precedes chemical reaction, the so called quasi-equilibrium theories of mass spectra, but at present the extent of the success is very limited and there are many instances where energy equilibrium is clearly not remotely approached. As the high initial energy of the activation process is progressively dissipated in the subsequent secondary reactions, species of rather lower energy, approaching that of species found in more 'normal' chemical reactions, begin to appear. Thus the very late stages of these radiolytic reactions may be used as a starting point for other chemical reactions involving species of moderate reactivity. For example electrical discharge is a convenient source of free atoms for kinetic studies in fast flow discharge systems (section 2.5b). At a time when all the charged species have disappeared from the discharge products by recombination processes free atoms, which take a longer time to recombine, remain and may exist for appreciable times as the only significant products of the discharge.

2.2d Chemical activation

By chemical activation is meant the use of the exothermicity of one chemical reaction to provide the activation for another. The exothermicity of a reaction in addition to its minimum energy of activation [i.e. the activation energy of the reverse step see (1.91)] must appear as excitation energy, either electronic, vibrational, rotational or *relative* translational, of the products. The actual distribution of this energy in addition to being limited by the usual conservation laws of momentum and angular momentum will depend on the detailed mechanics of the chemical reaction. Electronic excitation of the products when followed by allowed radiational transitions to lower states gives rise to the phenomenon of chemiluminescence and is, comparatively speaking, rare. In the case of combination reactions conservation of momentum prevents any excess energy appearing as translational motion, and so it must all go into internal degrees of freedom, usually as vibrational and rotational motions of the product. Rotational energy is relatively rapidly equilibrated by collisions. The vibrational energy however may remain in the product molecule sufficiently long for internal redistribution to occur which leads to decomposition or isomerization of the product by a route which is different from the reverse of that by which it was formed. Provided the energy of activation of the reverse of the first reaction exceeds the energy needed for activation of the second, chemical activation will occur. Alternatively the vibrationally excited product may transfer sufficient of its excess energy to a second species during a collision so as to activate this and cause it to undergo further reaction. If this latter reaction also produces a vibrationally excited product which can continue the process and so on indefinitely we have a process known as a vibrational energy chain. In this the vibrational energy of excitation is passed on repeatedly from reactant to product without thermalization occurring. Such vibrational energy chains were at one time supposed to be more common than is now thought to be so. No examples in thermally activated reactions are known but such chains may occur in some high energy U.V. photolyses, e.g. that of ozone, to some extent. In non-combinative reactions much of the energy released may appear as translational motion of the products relative to their centre of mass. Nevertheless some excess energy may remain in internal modes and so influence the subsequent reactions of the products. Transition to lower vibrational states accompanied by photon emmission leads to infrared chemiluminescence, the study of which can yield information concerning the distribution of energy in the products, (see pp. 75 and 129).

If an initial addition reaction, for example the recombination of free radicals or the addition of radicals and biradicals to unsaturated molecules,

requires little or no activation then the distribution of the total energy of
the adduct will be the Maxwell–Boltzmann energy distribution of the
reactants shifted to higher energies by an amount equal to the energy
liberated in the reaction provided that the probability of reaction is not
affected by the energy of the reactants. In many instances this seems to be
the case so that a relatively narrow energy spectrum of the products
results, if the temperature of the reactants is low and the exothermicity
is appreciable. Chemical activation thus provides a method of producing
activated molecules with rather precisely defined energies, and of studying
their rates of reaction. By using different reactants to produce the same
activated product with different amounts of energy, for example by isotopic substitution of D for H, the variation of the rate constant of the
subsequent reaction with total internal energy can be investigated (see
p. 120). The rates of reaction of chemically activated species are usually
determined by comparison with the rate of collisional stabilization of the
excited molecules, by measuring the ratios of the yields of the final products. The rate of stabilization is usually taken to be proportional to the
kinetic theory collision rate (section 3.2) and may be varied in a systematic
manner by varying the gas pressure.

    Chemical activation systems must by definition be complex, at least two
chemical processes, the production of the activated molecule and its
subsequent reaction, and one 'physical' process that of deactivation by
collisional energy transfer, must occur. Usually more processes than these
are involved since the initial highly reactive reactants must be produced,
usually by a complex process, and also more than one mode of reaction
of the excited molecule is usually possible and these themselves may involve subsequent reaction steps. The interpretation of the results of
chemical activation is therefore often complex and rarely unambiguous.

2.3 METHODS OF ANALYSIS

Having discussed the methods by which a chemical reaction can be caused
to occur we will now consider some of the more important techniques by
which the progress of the reaction can be followed. This involves studying
the variation due to the reaction of the number of molecules of reactants,
and intermediate and final products with time [see equation (1.9)]. Some
form of quantitative chemical analysis is therefore essential. For stable
species that can exist under normal laboratory conditions for some hours
the technique *par excellence* is gas-chromatography and to a lesser extent
mass-spectrometry. The latter technique is also extremely useful for the
analysis of less stable species such as free radicals and ions provided they
can exist for times of the order of 10^{-5} sec in a collision free system. Some

other techniques for measuring the concentrations of such unstable intermediates are also described briefly below. The list is by no means exhaustive but includes the more important analytical techniques that have been used in gas kinetics in recent years.

2.3a Gas-chromatography

The impact of gas-chromatographic analysis on the study of conventional thermally, photochemically and chemically activated reactions over the last 15 years or so has been enormous. It has virtually superseded all other means of quantitative analysis previously used, though for qualitative

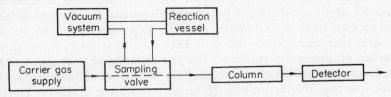

FIGURE 2.1   Block diagram of a gas-chromatograph.

analysis supplementary techniques are often needed. Not only has it made possible the study of reactions which could not be attempted before, but it has shown that much of the data obtained by older less reliable methods such as total pressure change measurement in closed systems, or low temperature distillation, etc., are erroneous or at least inaccurate. In the majority of cases therefore analytical kinetic studies made prior to the advent of gas-chromatography must be treated with some reserve and many of these systems have been or need to be re-investigated.

In its most usual form a gas-chromatographic unit has the outline shown in the block diagram of Figure 2.1. Small discrete gas samples to be analysed are taken from the reaction mixture either directly, or after quenching and preliminary separation, by a sampling valve and injected into the flow of the inert carrier gas which passes into the chromatographic column. In the column the sample material is partitioned between the moving gas phase and a stationary liquid or solid phase (or both). Substances with different partition coefficients spend on average different times in the stationary phase (but the same time in the moving gas phase) and hence emerge from the column after different total residence times. Their concentrations are separately measured by the detector as they emerge from the column. Very high separating efficiencies are relatively easily obtained, for example 10^4 to 10^5 plates is common (one plate corresponds to one perfect equilibration between the two phases followed by complete separation of the two phases by relative motion). Analysis times can be as low as a few seconds in favourable cases and even difficult separations take only a few minutes. Very sensitive detectors are available which can easily measure accurately (to better than 1%) the composition of samples of 10^{-10} moles. This corresponds to 1 cc of gas at a pressure of about 10^{-2} torr at room temperature. In favourable cases suitable detectors may improve on this figure by several orders of magnitude. Thus small gas samples may be taken from closed reaction systems without appreciably altering the concentrations thereby, and if the total pressures are in the normally convenient range of say 10–100 torr quite minor components can be easily measured. The method is very general, virtually all compounds are susceptible to analysis by this technique. Compounds differing only very slightly, e.g. isomeric forms are readily distinguished and so their interconversion and other reactions can be studied. As will be seen later such simple reactions have important implications in the theories of elementary reactions. For the quantitative analysis of stable products therefore gas-chromatography is without doubt the most useful technique available at present. However it is no use at all for reactive intermediate products such as atoms and free radicals, though intermediates of moderate stability may be analysed. There are theoretical reasons for supposing that gas-chromatographic analysis times cannot be reduced below about 1 to 10 milliseconds so that it is unlikely ever to prove of use for analysing species with shorter lives than this. This theoretical limit has not been approached at present, analysis times of $0 \cdot 1$ to 1 second being the best achieved so far. A very general qualitative and quantitative technique that will deal with such short-lived species as well as normal stable compounds is mass-spectrometry.

2.3b Mass-Spectrometry

A mass-spectrometer consists essentially of the following components; a source region in which the molecules under study are ionized (if they are not already in the form of ions) usually to positive ions; an analysing region in which ions of different mass are separated either spatially e.g. by the action of electric or magnetic fields or temporally, e.g. by a drift tube as in time-of-flight instruments; and a detecting system in which the current due to the passage through the instrument of the ions of various masses is measured, the ions of different masses being brought in turn for measurement onto the detector by some form of scanning in the case of spatially separated ion beams. Instruments vary widely in type from those with very high resolving power capable of separating ions with very small mass differences (e.g. 1 part in 10^5) and generally low scanning speeds (e.g. about 10 sec per decade of mass) to instruments such as time-of-flight spectrometers of relatively low resolution (e.g. 1 in 10^2-10^3) but very fast scan speeds (e.g. 10^4-10^5 spectra per second). Movement of the ions through the instrument must clearly take place under molecular flow conditions (see section 2.4b) to avoid the scattering effects of collisions. Pressures in the analysing section around 10^{-7} to 10^{-8} torr are normal, while those in the ionizing region are usually rather higher up to about 10^{-4} torr maximum. The ionization of the sample is usually achieved by electron, or sometimes photon, impact and measurement of the minimum electron (or photon) energy, the appearance potential, that will just cause a particular ionization process to occur can be made in this way.

For the chemical analysis of reaction mixtures of stable species mass-spectrometry has largely been superseded by gas-chromatography. This is due to the complex nature of mass-spectra, which arises since nearly every possible fragment ion is formed to some extent from each molecule present in the sample mixture, compared to the simple nature of chromatograms where each component yields only a single peak. The exception to this generalization lies in isotopic analysis of stable molecules. This is a field in which gas-chromatography has had only very limited success, most isotopic analysis still relies on mass-spectrometry. Isotopic analysis is useful in two main ways in kinetics. Firstly, isotopic labelling of reactant molecules provides a means of identifying the mechanistic pathways by which they are converted into products. For example isotopic mixing in the products can suggest the presence of free radical intermediates in a reaction which might otherwise appear to be molecular (e.g. see p. 194). More specifically the labelling of particular atoms in a molecule provides a means of following their fate in the reaction. Secondly isotopic substitution can yield information of direct interest to the theories of elementary reactions. The effect of isotopic substitution on the specific rate constant is called the kinetic isotope effect and can be used for example to indicate to what extent the substituted atom is involved in the nuclear motions which lead to reaction. Kinetic isotope effects may also be used to detect the occurrence of quantum mechanical 'tunnelling' in which an atom passes through rather than over a potential energy barrier, or to detect such features of potential energy surfaces (see section 3.7) as a depression near the top of the barrier.

A number of features make mass-spectrometry very suitable for the study of unstable reactive intermediate reaction products such as ions, atoms and free radicals. The low pressures existing in the instruments allow the species to travel in virtually collision-free paths during analysis so that loss due to reactive collisions with other gas-phase species or with the walls of the instrument are minimized. Also the time of flight through the instrument is very short, typically 10^{-5} sec or less, so that species with very short intrinsic lifetimes can be studied. Ions produced in external reaction systems such as flames, explosions, electric discharges, shock tubes, photolyses and radiolyses can be introduced by collision-free molecular beam sampling systems into the mass-spectrometer for analysis and measurement. Free radicals similarly can be introduced and their presence detected by the fact that the appearance potential for the radical ion obtained directly from the radical is lower than that of the same ion when produced from a molecule by an amount equal to the energy required to produce the radical from the molecule by dissociation. However to distinguish in this way between molecules and free radicals does require a relatively high proportion of free radicals to be present in the sample, say about 1%. Such high relative radical concentrations imply very fast reaction rates. Either steady state fast flow systems must therefore be used (sections 2.4b and 2.5a, b and c) or else a mass-spectrometer with a very fast scan speed is needed to follow the very rapid changes in concentration. Free radicals produced in flames, explosions, electric discharges, shock tubes, flash photolysis, etc., have been studied in this way. Relatively few rate constants for radical reactions have as yet been measured but much qualitative information about free radical intermediates has been obtained.

In addition to the uses so far outlined in which mass-spectrometry is used to study reaction systems which are external to the instrument, mass-spectrometers can clearly be used to study the kinetics of the processes occurring within them. Owing to the short time that species take to pass through the instrument such processes are clearly necessarily very fast. Typical first-order rate constants must exceed about 10^5 sec^{-1} for the reaction to be observed. Three main types of reaction may be studied. Firstly the kinetics of the reaction by which the ions are produced may be investigated. This includes such reactions as ionization by electron impact and photo-ionization. As discussed in section 2.2c such processes obey the Franck–Condon principle and produce non-equilibrium energy distributions in the product ions. Secondly the subsequent unimolecular decomposition reactions of the ions can be studied. These processes may or may not precede equilibration of the internal energy. The kinetic energies of the product ions may be measured relatively easily, e.g. by applying a retarding electric field, but little information about the internal energy, either its magnitude or distribution, is obtainable from conventional mass-spectrometric studies. Thirdly by increasing the pressure of gas in the ion source region of the spectrometer while maintaining the analysing section at normal low pressures the kinetics of ion-molecule reactions may be studied. The products of these reactions are readily distinguished from those of other processes since the intensities of the ion beams produced will be proportional to the *square* of the pressure in the ion chamber rather than directly proportional to it (see section 3.4). Much information on ion-molecule reactions has been obtained in this way and used to supplement data from more conventional radio-chemical experiments.

2.3c Some other methods for measuring free radical concentrations

In view of the importance of free radicals as reactive intermediates in many complex gas-phase reactions we will very briefly summarize some of the other techniques that have been used to analyse for free radical concentrations. The main disadvantage of all these techniques has been lack of adequate sensitivity. Special experimental systems containing relatively high radical concentrations have to be used in conjunction with these techniques. Some of these systems are described in more detail in section 2.5.

(*i*) *Metallic mirror removal*

This was one of the earliest techniques for detecting organic free radicals. These radicals will react with metals such as lead, zinc, antimony, etc., to form volatile organometallic compounds. The rate of loss of solid metal can be determined to give a semi-quantitative measure of the free radical concentration in the gas phase far from the surface. The technique is very inaccurate, non-specific and interferes grossly with the system under study. It is mainly of historical interest.

(*ii*) *Calorimetric probes*

When free radicals recombine on a catalytic surface the heat liberated may be measured and can be used to measure the concentration of free radicals far from the catalytic surface. Although more quantitative than mirror-removal it is also non-specific and considerably alters the quantity that is to be measured.

(*iii*) *Conversion of para to ortho hydrogen*

Any paramagnetic substance will catalyse the reversal of nuclear spin. By measuring the rate of conversion of para to ortho hydrogen caused by the presence of paramagnetic free radicals a measure of the total concentration of such species can be obtained. It is non-specific.

(*iv*) *Isotopic mixing*

In the presence of free radicals isotopic mixing reactions often become rapid and can be used to detect and measure the free radical concentration.

For example hydrogen atoms will convert D_2 to HD by the rapid exchange reaction

$$H + D_2 \rightarrow HD + D \qquad [2.1]$$

The rate constant of this reaction is known so that measurement of the rate of exchange yields a value for the concentration of hydrogen atoms. The technique is non-specific since almost any free radical will cause exchange and unambiguous quantitative results are difficult to obtain.

(v) Optical spectroscopy

Emission spectroscopy can detect excited species only, for example, electronically excited radicals in chemiluminescent reactions occurring in flames, explosions and electric discharges. It is completely specific of course, and can yield useful information about the structure of the emitting species. Relative concentrations of excited species can be obtained but it is difficult to get absolute values. It gives no information at all about the unexcited species which are usually present in vast excess compared to the species capable of emission. Absorption spectroscopy can be applied to all non-excited species which absorb in the accessible region of the spectrum. Relatively high concentrations however are needed to get measurable absorption (see, however, p. 83). Measurement of the extinction coefficient is often difficult so that often relative concentrations only can be obtained. Many species of great chemical interest do not absorb in a convenient region of the spectrum. Nevertheless much qualitative information has been obtained by using absorption spectroscopy in conjunction with special techniques for generating very high concentrations of radicals such as flash photolysis (section 2.5e).

(vi) Titration with stable species

Chemical reaction with a stable substance can be used to measure free radical concentrations in flowing gas streams such as exist in flames (section 2.5a) and flow discharge tubes (section 2.5b). If the free radical whose concentration is to be measured undergoes a very fast quantitative reaction with a stable species, for example NO, then this substance may be added to the gas stream in increasing amounts until the free radical is quantitatively destroyed. The added flow of the stable substance is then equal to the flow of the free radical and can be easily measured. From the known flow conditions the radical concentration can then be calculated from its measured flow rate. The end point of the titration can be determined by any of the other techniques described in this section for example by mass-spectrometry, the use of calorimetric probes, electron-spin-resonance spectroscopy, etc., but in some cases it is particularly simple to observe when chemiluminescent reactions are involved since the end point may be accompanied by an extinction or change in colour of the luminescence and can be observed visually or photoelectrically. For example oxygen atoms may be titrated with nitrogen dioxide. The quantitative reaction

$$O + NO_2 \rightarrow NO + O_2 \qquad [2.2]$$

will extinguish the yellow-green air afterglow emission due to the chemiluminescent reaction

$$O + NO + M \rightarrow NO_2^* + M \qquad [2.3]$$
$$\downarrow$$
$$NO_2 + h\nu$$

exactly at the equivalence point. Similarly hydrogen and nitrogen atoms can be titrated with nitric oxide, by the quantitative reactions

$$H + NO + M \rightarrow HNO^* + M \qquad [2.4]$$

$$N + NO \rightarrow N_2 + O \qquad [2.5]$$

High radical concentrations, for example partial pressures around 10^{-3} torr, are needed for this technique to be employed. A technique that has for a long time promised to enable identification and measurement of concentration of atoms and free radicals in 'normal' slow reactions where these concentrations are many orders of magnitude lower than the above figure is electron paramagnetic resonance spectroscopy, e.p.r.

2.3d Electron paramagnetic resonance

Free radicals, in addition to some less reactive molecular species have unpaired electron spins. This in general results in the species being paramagnetic. Exceptions to this rule occur when the spin and orbital contributions to the total magnetic moment cancel as for example in the ground state of nitric oxide. This is a $^2\Pi_{\frac{1}{2}}$ state and follows Hundt's coupling case (a) where both L and S are strongly coupled to the molecular axis. Since $\Lambda + 2\Sigma = 0$, this state has no resultant magnetic moment is consequently diamagnetic. The $^2\Pi_{\frac{3}{2}}$ state of nitric oxide however has $\Lambda + 2\Sigma = 2$ and hence is paramagnetic. The separation between this state and the ground state is small being of the order of magnitude of kT at normal temperatures so that the upper state is appreciably populated and nitric oxide gas is therefore paramagnetic at room temperature, but the paramagnetism is markedly temperature dependent. In intermediate coupling cases $^2\Pi_{\frac{1}{2}}$ states may be paramagnetic in the higher rotational levels as occurs for example in the case of the hydroxyl radical OH·. As a general rule therefore the presence of unpaired electron spins leads to paramagnetism and this is susceptible to observation and quantitative measurement. Direct measurement of the paramagnetism for example by Gouy balance is comparatively insensitive and of little value in measuring the minute concentrations of free radicals that exist in most systems of kinetic interest. However electron paramagnetic resonance, or electron spin resonance as the technique is sometimes described, has a much greater sensitivity than static susceptibility measurements and also gives specific information about the structure of the paramagnetic species present.

In this technique a magnetic field is applied to the paramagnetic gas thus causing Zeeman splitting of the energy levels. Transitions between these levels are then induced by either the oscillating magnetic or electric fields of microwave radiation and the resulting absorption of the microwave power recorded. By modulating the magnetic field (by employing an electromagnet carrying a direct current with a small A.C. component) a range of the spectrum can be traced out. This field variation is more convenient in practice than varying the frequency of the microwave radiation. Both magnetic-dipole and electric-dipole e.p.r. spectra may be obtained in suitable cases using suitable design for the absorption cavity. Magnetic-dipole transitions are more general since they do not require the substance to possess a permanent electric dipole.

In the case of atoms the spectra observed are quite simple. The transitions between the Zeeman levels are not complicated by rotational angular momentum of the species. Nuclear spin coupling to the electronic spin leads to hyperfine splitting of the levels but the number of lines observed is never very large. Hence high intensity is obtained in each line and the technique has moderate sensitivity for atoms. For example concentrations of O, H and N around partial pressures of 10^{-3} torr have been measured in fast flow discharges etc. However for more complicated species such as diatomic and polyatomic radicals the spectra are very much more complex. Coupling of the electronic angular momentum to the rotational motion of the molecule in general results in many lines appearing from different rotational states. This complexity of the spectrum means that the intensity in any one line is reduced so that the sensitivity tends to be much lower than that for atoms. In addition at total gas pressures above a few torr broadening of the lines takes place causing the spectrum to be 'smeared out' into a single broad absorption band with consequent further loss in sensitivity. For example the stable radical NF_2 gives a single broad absorption which is just measurable with radical concentrations of about 1 torr. Thus not only must the radical concentration be high but the total molecular concentration must be fairly low, i.e. a high percentage of radicals must be present which implies a very high reaction rate. The e.p.r. spectro-meter can be calibrated for quantitative measurements of radical concen-trations using molecular oxygen. The pressures of oxygen needed are in the region of 0·1 to 1 torr, i.e. some 10^2–10^3 times higher than that of atoms.

In some cases this lack of sensitivity may be overcome by using electric-dipole transitions. For example, the hydroxyl radical yields a complex spectrum due to rotational coupling as well as Λ doublet splitting and hyperfine splitting, but since the radical possesses a permanent electric dipole moment, electric-dipole transitions may be observed. The intrinsic intensity ratio for electric and magnetic-dipole transitions is $(\mu_e/\mu_0)^2$, where μ_e is the electric dipole moment and μ_0 is the Bohr magneton. For a typical electric dipole moment of $0·1 \times 10^{-18}$ e.s.u. this ratio is about 10^2. Thus about a factor of one hundred in sensitivity is gained by using electric-dipole transitions and so radicals of this type can be measured in fast reaction systems. For example the concentrations of $OH°$ produced by discharging water vapour in a fast flow system have been measured. Partial pressures of OH˙ as low as 10^{-7} torr have been measured.

Much work using e.p.r. has been directed to studies of the *structure* of free radicals produced in reacting systems by trapping them rapidly into a cold solid inert matrix followed by e.p.r. observation at leisure. Such techniques give no quantitative measure of the concentrations of the radicals present in the actual reaction system.

In theory concentrations in the region of 10^{10} spins per cc of gas, i.e. about 10^{-7} torr partial pressure of radicals should be measurable by e.p.r. In practice as outline above this sensitivity has not been approached to date except in favourable cases. A technique that does approach this sensitivity will now be described.

2.3e Optical pumping

In this technique atoms, such as those of rubidium, are spin polarized in a magnetic field by absorption of circularly polarized light. The light transmission gives a measure of the degree of spin orientation. These 'pumped' atoms lose orientation in collisions with the walls of the con-tainer or with inert gas molecules and by electron exchange with species containing unpolarized electrons. The presence of atoms or free radicals with unpaired electrons in the optically pumped gas will by virtue of electron exchange decrease the spin polarization of the pumped atoms and decrease the spin-lattice relaxation time (that is it will make spin depolariz-ation faster). If the atoms whose concentration is to be measured are in S states, for example, H, N, P, they will also acquire spin polarization by this electron exchange with the pumped (Rb) atoms. Application of a radio frequency magnetic field at the resonance frequency of the S-state atoms will depolarize them and hence reduce the polarization of the pumped atoms. The resulting decrease in the transmission of the polarized light can be measured. Thus S-state atoms can be identified and their concentrations can be distinguished from the sum effect of all other atoms and free radicals. The technique is therefore partially specific. Spin orientation is only possible for S-state atoms since other species have other angular momentum contributions apart from their electron spin. Thus non-S-state atoms have electronic orbital angular momentum while diatomic and polyatomic free radicals have molecular rotational angular momentum. The magnetic moment associated with this angular momentum produces a magnetic field acting on the electron spin which will change at almost every collision and so spin disorientation will be too rapid for any appreciable spin polarization to occur. Oxygen and other triplet states have spin–spin interaction which similarly prevents polarization. Although this technique is therefore not of completely general applicability it is extremely sensitive. Partial pressures of free radicals of about 10^{-8} torr can be measured. This means that radical concentrations in normal 'slow' reactions can be measured, for example, the concentrations of hydrogen atoms and other radicals can separately be measured in the steady state photolysis of ethane using a photolytic light source of 'normal' intensity (not a high energy flash). An alternative and more general method would be to orientate the nuclear spin of ^{3}He by collisions with optically pumped Rb and then to use this gas as a detector for free radicals but this technique would involve a loss of sensitivity of about a factor of 10^6 from the above figure.

2.4   GENERAL TECHNIQUES

Chemical reactions may be carried out either in closed or open systems. The closed system may not be of constant volume but by far the simplest case kinetically and the one most often used is the closed constant-volume or static system as it is usually called. Many chemical reactions occur too rapidly to be studied conveniently in such a system and in these cases an open system may be used in which reacting gases flow through a reactor or reaction region at high velocity so that changes in time become spread over appreciable distances along the flow. Such open systems for gas reactions operate under non-constant-volume conditions since, owing to the compressible nature of gases, the volume of any reacting element of gas will vary along the stream even if the temperature remains constant. This latter condition is often also not achieved since such rapid reactions as need to be studied in flow systems are often highly exothermic. As a result of the high reaction rate, cooling cannot be sufficiently rapid to prevent large temperature rises unless the system is highly dilute, i.e. unless an excess of an inert gas is present. Specific examples of these types of system will be considered in section 2.5 but first we will outline some of the experimental principles of static and flow systems.

2.4a Static systems

A static system consists essentially of a reaction vessel, often spherical or cylindrical in shape and usually made of glass, quartz or metal, situated in a thermostatically controlled oven of uniform temperature and connected to a vacuum system. Before use the vessel is evacuated to as low a pressure as possible (say less than 10^{-5} torr) and the reactant or mixture of reactants is then admitted to a measured pressure. If a mixture of reactants is to be used, this may be made up in a mixing vessel before being admitted to the reaction vessel or, alternatively, the reaction vessel may be constructed so that mixing can take place during the admission of the reactants. The time taken for the gases to flow into the vessel, reach the temperature of the oven and, if necessary, mix must obviously be much shorter than the time over which it is desired to study the reaction. This time for equilibration is usually about one second or less in properly designed systems so that reaction times of about 10 seconds upwards may be used without serious error from this cause. This places a limit on the speed of the reactions that may be studied in this way.

The course of the reaction may be followed in a number of ways. If the reaction results in a change in the number of molecules then it will be accompanied by a change in pressure in a closed constant-volume system. A much used method has been to follow this pressure change with a suitable gauge such as a mercury manometer or a recording pressure transducer. This method only measures one parameter of the system, the total number of molecules present, and so for complex reactions is of little use by itself. Much work on complex reactions has been done in the past using this technique alone which has lead to erroneous conclusions and the method has fallen somewhat into disrepute as a result. Analysis by one of the techniques described above is far more satisfactory, and may be carried out in several ways. Firstly, with suitable apparatus continuous monitoring of reactant and product concentrations can be achieved either *in situ*, e.g. by spectroscopy, or externally by having a very small continuous leak of the reaction mixture from the vessel into the analytical system. This leak must, of course, be sufficiently small not to alter the concentrations in the vessel appreciably during the time for which the reaction is studied. This requires a very sensitive analytical technique, such as mass-spectrometry. Alternatively, if continuous analysis is not possible, small samples may be removed from the reaction mixture at known times for analysis by a technique such as gas-chromatography. The samples must be very small compared to the total amount of reaction mixture so that the analytical technique must be very sensitive. If these sensitivity demands cannot be satisfied the whole of the reacting mixture may be removed from the vessel for analysis. To obtain information about the time-variation of the concentrations then requires a series of many experiments using identical conditions in which the reaction is stopped after various times.

The static system has the advantage that the conditions under which reaction occurs are as well defined as possible. The effect of the reaction vessel wall on the reaction, if any, can be found by changing the material of the surface and altering the surface area to volume ratio of the vessel. Corrections due to 'dead space', i.e. reaction volume outside the thermostatted region are relatively simple. Temperature and concentration gradients if they exist are fairly amenable to calculation, as also are any time lags in temperature equilibration or filling. Results of higher accuracy are obtained from these systems than from any other type. The main disadvantages of the technique are the relative slowness of the reactions that can be studied and the comparatively small amounts of material available for analysis. Both these disadvantages are overcome to some extent by the use of flow systems but only at the expense of losing much of the simplicity of the static system.

A compromise solution is the capacity-flow technique in which the reacting volume is efficiently stirred to obtain approximately homogeneous conditions in the reactor. Reactant flows into the vessel and the mixture is removed at a steady rate so that a steady state is set up in the reactor. Provided the stirring is sufficiently fast, the composition of the mixture in the vessel is uniform in space and steady in time. Thus, many of the difficulties inherent in normal flow systems (see below) are partially eliminated. The technique requires that the rate of stirring flow be much faster than the rate of reaction or diffusion. This limits the speed of reaction that may be studied.

2.4b Flow systems

The flow of compressible fluids is itself a highly complex phenomenon. If the fluid is also undergoing chemical reaction the situation, in general, becomes impossibly complex to deal with exactly. Under certain conditions some simplification is possible. Considering the flow of a gas in a cylindrical pipe the types of flow observed fall into four main regions, molecular, viscous (or streamlined), turbulent and supersonic flow. These may be distinguished by the following criteria. Molecular flow, where the gas moves as individual molecules rather than as a continuous fluid, occurs if the mean free path of the gas molecules between collisions is greater than the dimensions of the apparatus. Since the mean free path λ depends on the molecular concentration (c) according to the Kinetic Theory result [see equation (3.32)],

$$\lambda = 1/\sqrt{2}\pi c\sigma^2 \tag{2.2}$$

where σ is the collision diameter, this condition may be expressed in terms of the gas pressure. As a rough practical guide taking air as an example the flow is molecular if the pressure $p < 500/d$ microns. Where one micron (μ) is 10^{-3} torr pressure and d is the diameter of the pipe in cm. Molecular beam flow systems described later operate in this region (where λ may be many metres). If the mean free path is less than d then the flow is not molecular but may be viscous (streamlined) provided that the Reynolds' number for the flow has an appropriate value. The Reynolds' number is defined as $\bar{u}\rho d/\eta$, where $\bar{u}$ is the mean linear gas velocity, ρ is the gas density and η its viscosity. For long cylindrical pipes the flow ceases to be viscous and becomes turbulent if the Reynolds' number exceeds about 10^3. Again, as a rough practical guide, this condition may be put in the form $Q/d < 2 \times 10^5$ for viscous flow, where Q is the flow rate in $1\mu sec^{-1}$. If the flow is viscous the flow rate through the pipe is given by Poiseuille's equation, and although the velocity distribution across the pipe is parabolic, the *average* residence time of the reacting gas within a tubular reactor can be taken approximately as being the volume of the reactor divided by the total flow rate, that is by assuming plug flow. Under these conditions, therefore, time is related to distance along the reactor in a fairly simple fashion and the system is reasonably well suited to kinetic measurements provided results of very great accuracy are not required. When the Reynolds' number exceeds the value given above the flow becomes turbulent and the situation is considerably changed. The flow pattern is no longer streamlined and becomes very sensitive to the exact shape of the pipes, etc., through which the gas flows. In particular, any rapid changes in diameter of the pipe may have a marked influence on the flow pattern. For reactors which consist of a wide tube with narrow inlet and outlet tubes at opposite ends, the large change in diameter between the inlet tube and the main reactor itself can cause the onset of non-viscous flow at Reynolds' numbers, within the reactor, as low as 10. The result may be 'channelling' of the gas flow through the reactor, in which a narrow fast moving stream of gas passes through the reactor from inlet to outlet without expanding to fill the whole volume of the vessel, most of which is occupied by almost stationary gas. Under these conditions there is no simple way of calculating the residence time for the reaction and the simple plug flow approximation will give highly erroneous results. There is evidence to show that some work that has been conducted in conventional kinetic flow systems suffers from this defect. Kinetic results obtained from such systems must be treated with some reserve. Before a flow system can be used to obtain results of any quantitative kinetic significance it is essential that the fluid dynamic characteristics of the flow be determined so that the relation between flow rate and the effective average residence time can be found. This is equally true for flow in the fourth region where flow velocities exceed the velocity of sound, i.e. the thermal molecular velocities. The gas flow may be made supersonic by passing an inert gas through a suitably shaped convergent-divergent nozzle. The reactants may then be injected into the flowing gas stream in the supersonic region. Owing to the considerable experimental difficulties, applications of this technique are limited to a few simple and very fast reactions.

Owing to the compressible nature of gases, reactions in flow systems take place under conditions of non-constant volume, i.e. the reactant concentration varies with time even if no chemical reaction occurs. The kinetics are complicated by this fact, and this may be further increased by any change in the number of molecules (giving a pressure change) or by temperature gradients caused by the flow or by the reaction. Detailed analysis of these effects is complicated and often impossible. The use of flow systems to obtain very accurate kinetic data for gas reactions is consequently limited. In the case of very fast reactions, flow systems may be the only way of studying the reaction and their use can be justified as 'Hobson's choice'. Some of these flow systems will now be considered in more detail.

2.5 SPECIAL TECHNIQUES FOR FAST REACTIONS

2.5a Stationary flames

These are some of the oldest known fast chemical reactions. They may be *diffusion* flames where the mixing of reactants by diffusion occurs or *premixed* flames where a previously prepared mixture of reactants is burnt. The characteristics of flames vary over wide ranges, particularly with changing pressure. At low pressures the flames are called *dilute* or *rarified* and the temperature rise in the reaction zone of the flame, which is relatively large since the mean free path is long [see (2.2)], is very small. At high pressures normal '*hot*' *flames* occur in which most of the chemical reaction takes place in a very thin reaction zone and for exothermic reactions is therefore accompanied by a very rapid rise of temperature. For example, temperature rises of two or three thousand degrees within distances along the flow of one mm are not uncommon.

Well known examples of dilute flames are the low pressure diffusion flame reactions of alkali metals with halogens or halogen containing compounds such as alkyl halides. These flames were much studied between the 1930's and 1950's, but have fallen somewhat into disfavour owing to the difficulties of obtaining accurate quantitative results. An inert gas at a pressure of about 1 torr is passed over the alkali metal, say sodium at about 250°C, to become saturated with metal vapour at a partial pressure of about 10^{-3} torr. This gas stream diffuses through a nozzle into a flowing stream of halide vapour. At very low flow rates the reaction zone in which the processes

$$RCl + Na \rightarrow R^{\cdot} + Na^{+}Cl^{-} \qquad [2.6]$$

or

$$Cl_2 + Na \rightarrow Cl + Na^{+}Cl^{-} \qquad [2.7]$$

for example, occur, is approximately spherical and can be made visible, if it is not self-luminescent, by illuminating it with a sodium resonance lamp. The diameter of the zone is measured and this together with the known reactant pressures enables the rate constant to be calculated provided the diffusion coefficient is known. For diffusive mixing the diffusion coefficient controls the relation between distance travelled by the metal atoms and time (and hence rate of reaction). In practice, the reaction zone is not exactly spherical and exact solution of the fluid dynamic equations is complicated. The diffusion coefficients are often not accurately known. These depend on the collision diameters and may therefore be affected by the possibility of reaction occurring on collision. An additional complication is that the alkali metal halide produced is a solid and it is not clear what effect the nucleation of this in the reaction zone has on the reaction rate. For these reasons the results obtained from studies of these flames are not regarded as very reliable. For other reactions dilute flames are often used because the large size of the reaction zone makes detailed analysis of the composition changes through the zone much easier than at high pressure where the reaction zone is much thinner and the spatial resolution of the reaction is much less.

Reactions occurring in hot flames burning at pressures near atmospheric or above are usually complex. However, these flames can be effectively used as high temperature reactors in which to study more simple reactions at temperatures up to 3000°K or so. In such studies flat, one-dimensional flames are usually used. In these flames premixed reactants are fed to a burner which consists of a very large number of fine capilliaries and is surrounded by a similar guard-ring flame to isolate the flame under study from the atmosphere. In this way a good approximation to one-dimensional gas flow is achieved and the reaction zone consists of a flat plane zone (perhaps about 0·1 mm thick) perpendicular to the gas flow. The variation of temperature and composition with distance along the flow can be measured and hence related to time. The fast chemical reactions which occur in these flames cause non-equilibrium energy distributions and often the electronic, vibrational and translational temperatures differ markedly. Chemiluminescence of the flames is also a consequence of this disequilibrium. Ionization (which is the limiting case of electronic excitation) also greatly exceeds the 'thermal' value. Examples of elementary reactions which can be studied in these flames are the processes which occur when metals such as Cu, Na or Li are added to the flames, e.g.

$$M + Cu + H \rightarrow CuH^* + M \qquad\qquad [2.8]$$

$$H + H + Na \rightarrow Na^*(^2P_{\frac{3}{2},\frac{1}{2}}) + H_2 \qquad\qquad [2.9]$$

$$Li + H_2O \rightleftharpoons LiOH + H \qquad\qquad [2.10]$$

These reactions can also be used to measure [H]. Chemi-ionization reactions such as,

$$Na + H_2O \rightarrow Na^+H_2O + e \qquad\qquad [2.11]$$

$$Na^+H_2O + M \rightarrow Na^+ + H_2O + M \qquad\qquad [2.12]$$

and,

$$CH + O \rightarrow CHO^+ + e \qquad\qquad [2.13]$$

can also be studied.

2.5b Fast flow discharge tubes

By operating a flow tube reaction system at relatively low pressures of the order of a few torr the gas density is sufficiently low that quite high flow velocities of the order of 10^3 cm sec^{-1} can be achieved at low Reynolds' numbers (about 10), see p. 71. The simplicity of viscous flow is obtained yet fast reactions may be studied with a time resolution of 10^{-3} sec or better. Atoms may be generated in the flowing gas by passage through an electric discharge produced, for example, in a microwave cavity. Downstream of the discharge the more unstable ionized species will have decayed leaving behind the relatively more stable uncharged atoms and molecules. Provided the pressure is low and the walls of the tube are suitably deactivated, the recombination of the atoms may be sufficiently slow that appreciable atom concentrations (e.g. 1% of the total gas) exist for considerable distances, e.g. a metre or more, downstream of the discharge.

Nitrogen, hydrogen, and oxygen atoms are easily obtained in this way by discharging the diatomic gases, provided small quantities of certain impurities are present in the gas. The action of these impurities is poorly understood but there is considerable empirical evidence that their only effect is to increase the efficiency of dissociation without appreciable contamination of the product. Reactions of the atoms produced in this way may be studied by adding other reactants to the flow downstream of the discharge. The progress of the reaction along the tube can be followed by the techniques described in section 2.3, for example, mass-spectrometry, calorimetric probes, electron paramagnetic resonance spectroscopy, chemiluminescence emission spectroscopy, absorption spectroscopy and chemical titration have been used. Infra-red chemiluminescence studies on reactions such as

$$H + HO_2^{\cdot} \rightarrow OH^{\cdot} + OH^{\cdot} \qquad [2.14]$$

$$H + NO_2 \rightarrow NO + OH^{\cdot} \qquad [2.15]$$

$$H + Cl_2 \rightarrow HCl + Cl \qquad [2.16]$$

have also been made in these systems and have given much information about the distribution of vibrational energy in the products of these reactions (see p. 129).

2.5c Crossed molecular beams

If the pressure of a flowing gas is reduced below 10^{-5} torr the mean free path exceeds several metres and molecular flow occurs in apparatus of normal dimensions. This flow may be collimated by suitable slits into a molecular beam of virtually non-colliding molecules whose translational velocity has an accurately known direction and a magnitude that is determined by the temperature of the source of the molecules. If two such beams are arranged to cross, for example, at right angles there is a small probability that collisions will occur between molecules of the two beams under very closely specified conditions. The result of these collisions may be elastic scattering, inelastic scattering (energy transfer) or reactive scattering (chemical reaction). The scattered molecules may be observed and their numbers measured by a suitable detector. This is mounted so as to be movable about a circle centred on the crossing point of the beams, in order to study the angular distribution of the products. The velocities of the incident molecules may be selected and the velocities of the scattered molecules measured by using rotating slotted discs or rotating helical grooves to stop all molecules except those with the right velocity to pass through the moving apertures. Thermal sources for the beam molecules yield molecules with translational energies up to about 0·5 eV (12 kcal mole^{-1}) while neutralization, by charge transfer, of ion beams yields conveniently molecules with translational energies greater than 10 eV (230 kcal mole^{-1}). The gap in available energies may be filled by use of supersonic beams. States differing in the orientation of their magnetic moment may be selected by using inhomogeneous magnetic fields. Rotational states may be separated by electrostatic quadrupole fields. Vibrational state selection is also possible. Flight times may be measured by using pulsed or modulated beams. Thus the microscopic details of collisions can be studied.

The crossed molecular beam experiment is the closest approximation yet achieved to the ideal experiment of gas kinetics in which two molecules in known quantum states with known relative velocity interact to form products in specified states with a collision cross-section that is measureable as a function of all the parameters of relative kinetic energy, impact parameter, orientation, internal energy states etc. In all other techniques only the total reaction cross-sections averaged over particular statistical distributions of these variables are observed. The molecular beam reactions are also free from the complications of competing processes and chemically reactive species may be studied owing to the dilute nature of the system. It is also clearly well suited to the study of ion-molecule reactions.

The main disadvantage of the method is the very high sensitivity required of the detectors owing to the low numbers of molecules to be measured. The highly sensitive surface ionization detector in which the product molecule is ionized on the surface of a metal filament, whose work function exceeds the ionization potential of the molecule, with nearly 100% efficiency is limited to species with low ionization potentials such as alkali metals and their compounds. Differential use of two such detectors with different metals which respond to different species can be used to distinguish between reactants and products for example, K and KBr in the reactions

$$K + RBr \rightarrow KBr + R^{\cdot} \qquad [2.17]$$

The use of electron impact to ionize beam molecules for mass-spectrometric detection, although allowing detection of all species has a very much lower efficiency of ionization and hence a very much lower sensitivity. Nevertheless, it has been used in conjunction with electron multiplier detectors to study molecules that do not contain alkali metals. For example,

$$H + H_2 \rightarrow H_2 + H \qquad [2.18]$$

has been studied in this way. To alleviate the detector sensitivity requirements less specific information about reactive collisions can be gained by using a single molecular beam and a target gas. In this case the reaction zone is not defined as a small spatial region nor is the relative velocity of the reactants so clearly specified, owing to the Maxwellian distribution of velocities of one of the reactants. The crossed molecular beam method is clearly preferable in principle, for no other technique can yield such detailed information about the dynamics of elementary chemical processes.

2.5d Shock tubes

The use of systems with stationary shock waves to study fast chemical reactions has already been mentioned (p. 72). A method of far more widespread use is that in which a moving shock wave is propagated down a tube containing the reaction mixture. Some of the interest shown in this technique undoubtedly stems from the importance of the chemistry of reactions in shock waves to the problems of supersonic flight. Nevertheless, much information of direct chemical interest has been gathered by this technique even though the conditions within and behind the shock front are far removed from those of more conventional thermal equilibrium systems.

The shock wave is usually generated by rupturing a diaphragm separating a high pressure 'driver' gas from the reaction mixture at a lower pressure. Using pressure ratios up to about one hundred enables shock velocities up to about Mach 10 to be obtained (10 times the speed of sound). The shock wave travels through the reaction mixture at supersonic velocities as an approximately plane wave provided the tube is of sufficiently large diameter. Upon striking the far end of the tube the shock may be reflected back through the mixture thus intensifying the effect of the incident shock. In the shock front the pressure, density and translational temperature of the gas jump almost discontinuously. The thickness of the transition region is of the order of the mean free path of the gas molecules (say about 10^{-5} cm at N.T.P.). Temperature rises of 10^4 °K and heating rates of about 10^2 °K per μsec can be obtained. Thus the shock tube provides a method of preparing a reactant gas mixture at high temperatures under virtually homogeneous conditions in a very short time so that very fast reactions may be studied. Expansion waves may be used similarly to obtain quenching rates of up to 10°K per μsec so that reactions can be both started and stopped quickly. Equilibration of the translational temperature in the shocked gas takes place in about 5 to 10 collisions depending on the shock velocity. Rotational equilibration of all gases other than H_2, D_2 and HD (which have large rotational quanta) follows sufficiently rapidly that the relaxation cannot usually be separated from that of translation. However vibrational relaxation (equilibration) and chemical reactions that have half-lives around 10^{-4} sec at 1 atmosphere can be distinguished as occurring behind the shock front. Thus not only normal chemical processes but also the mechanisms of vibrational energy transfer and excitation and ionization processes can be studied. At high temperatures (10^4 °K) chemical reactions such as the dissociation of a diatomic molecule and its reverse may proceed sufficiently fast to grossly disturb the Boltzmann distribution of vibrational energy. Such reactions under non-equilibrium conditions have been much studied by this method. For slower reactions with moderate activation energies, say >20 kcal mole^{-1}, the shock tube may be used in a manner analogous to the conventional static system. The gas mixture is heated rapidly by a compression shock and then after a known time delay it is quenched by an expansion wave. The reaction products in the quenched gas can then be analysed by gas-chromatography or mass-spectrometry, etc.

One of the principal disadvantages of the shock tube technique is the difficulty of measuring the reaction temperature with any precision. Under non-equilibrium conditions the significance of the 'temperature' is questionable anyway but even for chemical processes occurring after all the thermal energy has relaxed to equilibrium distributions the temperature can rarely be specified to closer than ±50°K. For very high temperatures this error may be small as a percentage of the absolute value. For moderate temperatures the difficulty of getting accurate data is sufficient that often the technique is used in a qualitative fashion only or else as a comparative method. A known reaction rate is used to monitor the temperature for a non-interacting unknown one. If only relative rates are needed the absolute temperature of reaction is relatively unimportant. Difficulties also arise in the use of techniques such as spectroscopy or mass-spectroscopy to follow events in and behind the shock front. The system must often be viewed through the cool boundary layer next to the wall of the shock tube or the sampling tube with consequent error. For reflected shocks heat loss to the end wall also is important and contributes further to the uncertainty of the temperature rise as measured by the shock velocity.

2.5e Flash photolysis

The use of steady light sources of relatively low intensity to initiate chemical reaction has been outlined in section 2.2b. Under normal photolysis conditions the rate of the primary photochemical reaction is low owing to the limited intensity of normal light sources. Hence in reactions which involve reactive intermediates such as atoms, radicals and excited molecules the concentrations of these intermediates will be very low and their presence may only be inferred either indirectly by use of steady state arguments (section 1.6) or else by such a highly sensitive technique as optical pumping (section 2.3e). However by using a very intense light source the rate of reaction and the concentrations of these reactive intermediates may be greatly increased thus making their direct observation possible by less sensitive but more general techniques such as absorption spectroscopy (section 2.3c) and mass-spectroscopy (section 2.3b). The sort of light intensity required for this purpose can only be obtained in short pulses or flashes. However this short flash duration is itself in some ways an advantage since the very fast subsequent reactions of the high concentrations of intermediates that are produced can then be followed by suitably rapid analytical techniques. The detailed progress of these reactions with time can thus be studied rather than in the steady state of normal photolysis. This very fast initiation thus enables very fast reactions to be studied.

The intense light flash is usually obtained by a pulsed electric discharge through a flash lamp containing an inert gas. The light emitted is mainly a continuum. Pulsed lasers and mercury discharges have also been used to obtain monochromatic flashes. The electrical energy required for the normal discharge flash lamp is stored prior to the flash by charging large condensers (about $100 \mu F$) up to high voltages (about 10^4 V). Typical flash energies ($\frac{1}{2}CV^2$) are thus around 10^4 joules of which about 15% emerges from the lamp as light in the photochemically useful region of the spectrum, (2000–4000Å). By suitable design of the electrical circuit, that is by reducing mutual and self inductance and resistance to a minimum, flash times as low as $1 \mu sec$ can be obtained, though often longer flashes are used (10–1000 μsec). Pulsed lasers may improve on this shortest time by a factor of 10^3 or so. With conventional flash lamps reactions with time constants down to about $5 \mu sec$ can be studied provided suitable analytical systems are used. The total actinic light from a normal sized flash lamp may therefore approach one Einstein per second (one Einstein equals one Avogadro's number of photons). If this were all absorbed by the reactant the amount of initial photochemical decomposition would clearly be very high, in some cases approaching 100% (for example I_2 vapour can be almost completely dissociated to atoms). To obtain efficient use of the light the reaction vessel and flash lamp are usually arranged to lie parallel and the two tubes are surrounded by a cylindrical reflector coated with magnesium oxide. Thus high concentrations of reactive species are made available for study. Alternatively the stable products left after all reaction has ceased can be studied by normal analytical methods. Owing to the high intermediate concentrations formed in flash experiments the final products are often very different from those of normal low intensity photolysis. For example radical–radical reactions are far more important in flash photolysis whereas radical–molecule reactions predominate in low intensity photolysis.

The most frequently used method of studying the intermediates produced in flash photolysis is that of kinetic absorption spectroscopy. The absorption spectrum of the reaction mixture is photographed at a known time interval after the main photolytic flash by using the light from a low intensity short duration spectroscopic flash. This is fired by a trigger pulse which is derived from the main light flash by way of a variable time delay circuit. By repeating the experiment with different time delays between the photolytic and spectroscopic flashes the variation in the absorption spectrum with time is obtained. For simple systems an alternative method is that of continuous absorption spectroscopy. The spectroscopic flash is replaced by a steady light source such as a tungsten filament lamp and the photographic plate is replaced by a photo-cell or photo-multiplier the output of which is displayed on a cathode-ray oscilloscope. In this way the variation in absorption at one particular wavelength with time is obtained from each experiment. This may be sufficient to obtain the variation of the concentration of one component of the reaction mixture with time, for example, if only one species present absorbs appreciably at the wavelength studied. Absorption spectroscopy although particularly suited to the flash technique has the limitation that many kinetically interesting species do not have absorption spectra in the conveniently accessible region or alternatively species showing continuous absorption may be present and obscure the spectra by overlapping. The application of infra-red and microwave spectroscopy can largely solve these difficulties but the experimental difficulties involved in the short-time-resolution applications of these at present are very great.

One of the major problems of all photolysis experiments is associated with the photochemical heating effect. That is to say the light energy absorbed by the system eventually becomes degraded in part to thermal energy which may result in a large temperature rise in the reaction mixture. In the case of flash photolysis the effect is particularly large since owing to the very short duration of the flash the heat lost to the surroundings by the reactant is negligible during the flash so that the flash is absorbed virtually adiabatically. Unless specific steps are taken to prevent it, the temperature of the reaction mixture rises very rapidly during and immediately after the flash. Typically the temperature may rise by about one thousand degrees in less than one millisecond. In some applications this effect may be turned to advantage since it affords a method of raising the temperature of a gas mixture in a rapid homogeneous fashion to high values. It may be compared to the use of shock tubes in this respect (section 2.5d). Homogeneous explosions in gas mixtures can be caused in this way and have been much studied by this technique. To use flash photolysis under virtually isothermal conditions this temperature rise must be prevented by adding a large excess of an inert, non-absorbing or reacting, gas to increase the heat capacity of the reaction mixture. In this way the temperature rise may be reduced to a few degrees at the expense of considerably diluting the system under study.

The flash technique has found much application in the production of reactive species such as free radicals and metastable electronic states for both structural and kinetic studies. Non-equilibrium vibrational, rotational and electronic energy distributions may be produced for energy transfer studies. Simple dissociation-recombination reactions, fluorescence and phosphorescence of aromatic hydrocarbons in inert glassy solids, complex explosion and combustion reactions are all examples of reactions which have been widely studied by this technique. Much of the kinetic information obtained is of a rather qualitative or semi-quantitative nature and often relates to conditions far removed from those of 'normal' conventional kinetics studies. Nevertheless when used in conjunction with thermal activation and other more conventional techniques it is a powerful tool for exploring many systems of kinetic interest. The development of monochromatic flash photolysis using pulsed lasers will facilitate measurement of quantum yields and enable excitation to specific states to be used.

2.5f Intermittent photolysis

The use of shock waves and intense light flashes described above are examples of studies of reaction rates under non-steady state conditions produced by a sudden, almost discontinuous, jump in the rate of activation (thermal and photolytic in these cases). An alternative way of studying non-steady state reactions is to employ intermittent activation. For example, in the case of photolytic activation the light may be switched on and off at regular intervals. This may be done conveniently for example either by having a steady light source situated behind a shutter such as a rotating disc with alternate opaque and transparent sectors or by using a fluorescent discharge tube powered by a square wave generator, of variable frequency. Intermittent thermal activation can be achieved by using sound waves to produce periodic (translational) temperature fluctuations by virtue of the periodic adiabatic compression and expansion that occurs in the wave. Intermittent radiolytic activation can be obtained by varying the distance between the radiation source and the reaction vessel in a periodic fashion. The principles of these techniques are all rather similar and will be illustrated here by describing intermittent photolysis or the rotating sector technique as it is often called.

If a complex reaction mechanism involves reactive intermediates then as discussed in section 1.6 the reaction will have an induction period (t_i) during which the concentrations of these intermediates rise towards steady state values, (see Figure 1.3c). Thus when the light is switched on the concentrations of these reactive intermediates will rise towards the steady state value with a time constant t_i and similarly when the light is turned off the concentrations will fall at a rate which is also governed by this same factor, see equations (1.120) and (1.130). If the rate at which the light is turned on and off is very slow so that the induction period t_i is very much less than the period of the complete light cycle, then the concentration of intermediates will reach the steady state value very soon after the light is turned on and will fall to zero very soon after the light is turned off. The rate of reaction, which depends on the concentration of reactive intermediates, will vary similarly. Suppose for the sake of argument that the periods of light and darkness are equal in length. Then the total amount of reaction over a large number of light cycles will simply be the same as the amount that would occur in steady state illumination for one half the total length of time. That is, the average rate of reaction over the whole period (including both light and dark periods), will be exactly one half the steady state rate of reaction. For unequal light and dark periods the corresponding factor is easily calculated similarly. Now suppose that the rate of rotation of the sector is made very fast indeed so that the period of the light cycle becomes very much shorter than the induction period. The intermediate concentration cannot rise and fall between zero and the steady state value with the rapidly varying light intensity. Instead a new effective steady state concentration of intermediate is set up (on which is superimposed a very small oscillation at the frequency of the light pulses) which corresponds to the time-averaged value of the light intensity. In the case of equal periods of light and darkness this will be exactly one half the intensity of the light during the 'light on' period. Thus the reaction effectively 'sees' a light source whose intensity is halved by the rotating sector. If the rate of reaction is proportional to the first power of the light intensity there is no difference between the effect of a light of intensity I on for half the time and a light of intensity $I/2$, so that the average rate of reaction will be independent of the rate of rotation of the sector. If however, as occurs in many reactions, the rate of reaction is proportional to some power of the light intensity other than the first the two cases will not give the same average rate and this rate will therefore be a function of the rotation speed of the sector. For example, it is a common feature of some complex reactions that the rate is proportional to $I^{\frac{1}{2}}$ (see section 4.3). Then at low rotation speed the average rate of reaction will be proportional to $\frac{1}{2} \times I^{\frac{1}{2}}$, while at high pulse rates the average rate will be proportional to $(I/2)^{\frac{1}{2}}$. Thus the ratio of the observed rate to the steady state rate will change from $0\cdot 5$ to $1/\sqrt{2} = 0\cdot 715$, as the sector speed is increased. The mid-way point between these two extremes occurs when the period of the light pulses is approximately equal to the induction period. More detailed mathematical analysis than the simple qualitative picture described here allows a quantitative relation between these two quantities at the mid-way point to be derived for any given reaction mechanism. Determination of the pulse rate at this mid-point thus enables the induction period to be calculated. Induction periods are related to the rate constants for the reactions by which the reactive intermediates are destroyed, see equation (1.130). Rate constants for many radical recombination reactions have been determined in this way, see section 4.3i for example.

The simple measurement of the effect of pulse rate on the average overall rate of reaction described above yields only one characteristic parameter of the reaction, the induction period. The method is therefore only applicable to reactions whose mechanisms are known from other studies, in addition to being limited to only those reactions whose rates are not directly proportional to the light intensity. A refinement of the technique that gives more information about the reaction and removes these restrictions has been developed by using absorption spectroscopy with modulated light sources (to obtain the very high sensitivity needed) to measure the concentrations of the species present in the reaction mixture under intermittent illumination. The average concentrations over the first halves of the periods of illumination are measured separately from those over the second halves and similarly for the dark periods. Thus information about the actual *shape* of the rise and fall of concentrations with time is obtained instead of simply its time constant. The phase angle differences between the illumination and the concentrations of the species present are thus measured and used to check possible mechanisms. These studies are intermediate between the simple rotating sector technique and the flash technique (in which the complete variation of composition during a single pulse is measured) but do not involve abnormally intense illumination.

SUGGESTIONS FOR FURTHER READING

Textbooks

General

E. F. Caldin, *Fast Reactions in Solution*, Blackwell, Oxford, 1964.

H. Melville and B. G. Gowenlock, *Experimental Methods in Gas Reactions*, Macmillan, London, 1964.

A. Weissberger, *Techniques of Organic Chemistry*, Vol. VIII, Parts I and II, Wiley, New York, 1961.

Special Topics

I. Amdur and G. G. Hammes, *Chemical Kinetics*, (for shock tubes and molecular beams), McGraw-Hill, New York, 1967.

P. G. Ashmore, F. S. Dainton and T. M. Sugden, *Photochemistry and Reaction Kinetics*, Cambridge University Press, Cambridge, 1967.

P. B. Ayscough, *Electron Spin Resonance in Chemistry*, Methuen, London, 1967.

A. G. Gaydon and H. G. Wolfhard, *Flames*, Chapman and Hall, London, 1953.

E. F. Greene and J. P. Toennies, *Chemical Reactions in Shock Waves*, Arnold, London, 1964.

J. H. Purnell, *Gas Chromatography*, Wiley, New York, 1962.

A. J. Robertson, *Mass Spectrometry*, Methuen, London, 1954.

Reviews

Thermal activation, *Quart. Rev.*, **11**, 87 (1957).

Photolytic activation, *Advan. Photochem.*, **1**, 1 (1963).

Radiolytic activation, *Progr. Reaction Kinetics*, **3**, 97 and 303 (1965). *Quart. Rev.*, **17**, 101 (1963); **20**, 153 (1966).

Chemical activation, *Quart. Rev.*, **18**, 122 (1964).

Mass-spectrometry, *Quart. Rev.*, **13**, 215 (1959).

Electron spin resonance, *Quart. Rev.*, **12**, 520 (1958).

Fast reactions, *Disc. Faraday Soc.*, **17**, (1954).

Atom reactions, *Progr. Reaction Kinetics*, **1**, 1 (1963); **3**, 63 (1965); **4**, 1 (1966). *Quart. Rev.*, **15**, 237 (1961).

Flames, *Ann. Rev. Phys. Chem.*, **13**, 369 (1962). *Quart. Rev.*, **4**, 1 (1950); **5**, 44 (1951); **17**, 243 (1963).

Flow-discharges, *Ann. Rept. Chem. Soc.*, **LXII**, 17 (1965).

Crossed molecular beams, *Advan. Chem. Phys.*, **X**, 319 (1966). *Ann. Rept. Chem. Soc.*, **LXII**, 39 (1965). *Quart. Rev.*, **20**, 465 (1966).

Shock tubes, *Ann. Rept. Chem. Soc.*, **LXII**, 63 (1965). *Ann. Rev. Phys. Chem.*, **16**, 245 (1965). *Quart. Rev.*, **14**, 46 (1960). *Sci.*, **141**, 867 (1963).

Flash photolysis, *Quart. Rev.*, **10**, 149 (1956).

Pulse radiolysis, *Progr. Reaction Kinetics*, **3**, 237 (1965).

CHAPTER THREE

THEORIES OF ELEMENTARY REACTIONS

3.1 INTRODUCTION

In this chapter are presented some of the attempts to predict the rates of
elementary reactions theoretically in terms of properties of the reacting
species that are either calculable, e.g. from fundamental quantum mechanical laws or experimentally measurable, preferably in non-reacting
systems. These theories fall into two main categories depending on the
model chosen to represent the elementary chemical act. The first of these
to be developed was the collision model in which attention is focussed
firstly on the initial state of a single elementary process when the molecules
that are to undergo reaction are still widely separated and not interacting.
By using the appropriate mechanical laws governing the molecular
motions an attempt is made to predict the course of single interactions
between the reacting molecules right through to the final state where the
molecules are again widely separated. Thus the probability of interactions
leading to chemical reaction can be predicted and by suitable statistical
averaging the overall rate of reaction in a large assembly of reactant
species is obtained. Unfortunately the approximations that had to be
made to enable this calculation to be carried through in the earliest
versions of this type of theory were so restricting that the results were of
very limited applicability. To partially overcome these difficulties the
second type of theory was developed in the middle 1930's and has been
called by many names such as activated complex theory, or absolute rate
theory. Here the title *Transition State Theory* (T.S.T.) will be used as being
a more accurate description of this type of approach. In transition state
theory attention is focussed mainly on the strongly interacting system of
molecules part-way between the initial state and the final state of a single
elementary chemical act and the reaction rate is formulated in terms of the
properties of this 'transition state'. The complex problem of calculating
the precise manner in which the system approaches and leaves this state is
circumvented, by assuming a quasi-equilibrium between the initial and
transition states. For this reason these theories are sometimes called
quasi-equilibrium theories. Transition state theory largely dominated the
field of reaction rate theory for twenty or thirty years after its initial
development but in recent times the more fundamental and detailed collision theory has advanced considerably both as a result of improvements

in computational methods and as a result of the more detailed experimental results obtained from such techniques as crossed molecular beams, discussed in Chapter 2.

Collision theories will be described first, beginning with the simplest of all theories the *Simple Collision Theory* (S.C.T.) and going on to more recent work that will be dealt with largely in a qualitative fashion owing to the mathematical complexity of these theories. The second part of the chapter will describe Transition State Theory.

COLLISION THEORIES

3.2   THE SIMPLE COLLISION THEORY OF BIMOLECULAR REACTIONS

In order for chemical reaction to occur between two gas molecules it is assumed necessary, firstly, for the molecules to 'collide' in the kinetic theory sense. The gas molecules are taken to be spherical and of a diameter σ, the collision diameter. Since the molecules are assumed to be

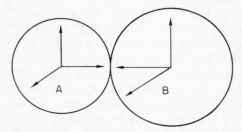

FIGURE 3.1   Translational velocity components for a bimolecular collision

infinitely hard and non-interacting except when in contact this collision diameter is constant and independent of the nature of the collision or of the property of the collisions (e.g. energy transfer, or momentum transfer or chemical reaction) that is being considered. Secondly in order for chemical reaction to result from such a collision it is assumed that the 'violence' of the collision must be sufficient. In other words, the relative kinetic energy of approach of the two spheres along the line of centres must exceed a critical minimum value say, ε_c. For real molecules rather than infinitely hard ones one can imagine that the closeness of approach, and the amount of distortion of the molecules on collision are determined largely by this part of the total energy of the molecules (Figure 3.1). Translational motion at right angles to the line of centres, internal vibrational motion, and rotational motion of the molecules would seem intuitively to be of lesser importance and so are ignored completely in the first instance.

All other conditions that may be necessary for reaction to occur are also ignored in the simplest theory. These factors will include any energy distribution or phase requirements for the internal molecular vibrations, any orientation of the molecules or parts of the molecules that may be necessary and the effects of any barriers to reaction due to steric blockage. An empirical correction term, P, called the 'steric factor' is then introduced to allow for the combined effect of these factors on the probability of reaction occurring when two molecules collide with sufficient violence. Since this term may represent the multiplication together of a large number of separate probabilities we may expect it to take a very wide range of values. It is not, as often stated, simply the fraction of collisions in which the molecules are suitably orientated in space. This simple orientation factor would never be expected to be a very small fraction (e.g. less than 10^{-2}) even for molecules of moderate complexity since it is unlikely that the orientation requirements would be so specific.

3.2a The total collision rate

The rate of collision between pairs of molecules, A and B, in a mixture of ideal gases A and B in thermal equilibrium may be calculated assuming that no chemical reaction occurs. Consider a single 'typical' A molecule moving with a speed $\bar{v}_{AB}$ in an arbitrary direction, where $\bar{v}_{AB}$ is the average magnitude of the relative velocity of the A and B molecules. In unit time it will collide with all B molecules whose centres lie in the cylindrical volume $\pi\sigma_{AB}^2\bar{v}_{AB}$ (see Figure 3.2), where

$$\sigma_{AB} = \tfrac{1}{2}(\sigma_A + \sigma_B) \qquad (3.1)$$

is the mean collision diameter of A and B, and can be determined experimentally by studying mass, momentum, or energy transfer, i.e. diffusion, viscosity, or thermal conductivity respectively, in the separate pure gases, since A and B are assumed to be hard spheres. The rate of collision of this single A molecule is,

$$\text{rate} = \pi\sigma_{AB}^2\bar{v}_{AB}C_B \text{ (collisions sec}^{-1}) \qquad (3.2)$$

where C_B is the concentration of B in molecules per cc. If there are C_A molecules of A per cc then the collision rate of all A molecules with B molecules per cc of mixture is

$$_2r = \pi\sigma_{AB}^2\bar{v}_{AB}C_AC_B \text{ (collisions sec}^{-1} \text{ cc}^{-1}) \qquad (3.3)$$

Strictly speaking both the speed and direction of A will change on every collision. However, by choosing only those segments of the path of the A molecule in which it *is* travelling with speed $\bar{v}_{AB}$ in a given direction and

4

joining them up to form a cylinder swept out in unit time, Figure 3.2 would be the result. Also, we should consider an arbitrary relative velocity v_{AB} and calculate the collision rate for this value before averaging over all velocities. However, since the final result for the collision rate is directly proportional to $\bar{v}_{AB}$, reversing the order of the averaging procedures does not alter the result and may be used for simplicity. More rigorous derivations of this result may be found in texts on *Kinetic Theory*.

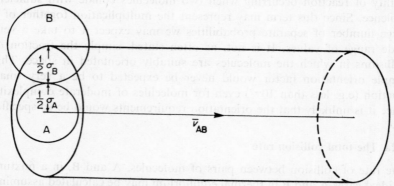

FIGURE 3.2   The collision rate of a 'typical' A molecule with B molecules.

The collision rate is often written,

$$_2r = Z_{AB}C_AC_B \tag{3.4}$$

where,

$$Z_{AB} = \pi\sigma_{AB}^2\bar{v}_{AB} \text{ (cc molecule}^{-1}\text{ sec}^{-1}) \tag{3.5}$$

Z_{AB} is called the collision frequency or collision number and is the collision rate per unit volume for unit concentrations of A and B. In a pure gas, A, the above formulae must be divided by two since each collision will have been counted twice if A and B are identical. Thus,

$$Z_{AA} = \tfrac{1}{2}\pi\sigma_A^2\bar{v}_{AA} \tag{3.6}$$

The value of $\bar{v}_{AB}$ given by the Maxwell–Boltzmann velocity distribution for an equilibrium system (no reaction) is,

$$\bar{v}_{AB} = \left(\frac{8kT}{\pi\mu}\right)^{\frac{1}{2}} \tag{3.7}$$

where μ is the reduced mass given by,

$$\frac{1}{\mu} = \frac{1}{m_A} + \frac{1}{m_B} \tag{3.8}$$

For a single gas,

$$\bar{v}_{AA} = \left(\frac{8kT}{\pi \frac{1}{2} m_A}\right)^{\frac{1}{2}} \tag{3.9}$$

$$= \sqrt{2}\bar{v}_A \tag{3.10}$$

where $\bar{v}_A$ is the mean molecular velocity.

3.2b The fraction of collisions of sufficient violence

The relative kinetic energy of approach along the line of centres in a collision of A and B may be written as,

$$\varepsilon = \frac{p^2}{2\mu} \tag{3.11}$$

$$= \frac{p_A^2}{2\mu} + \frac{p_B^2}{2\mu} \quad \text{(say)} \tag{3.12}$$

where p is the momentum along the line of centres relative to the centre of mass. This energy ε may be distributed in any way between A and B, for example molecule A might be moving fast and B almost stationary or vice versa or any other combination but provided the total relative kinetic energy adds up to the critical amount ε_c, or more, reaction may occur. This is expressed by saying that the critical energy may be distributed over two squared terms or two degrees of freedom, since two is the maximum number of *independent* component terms involving the square of a coordinate or momentum into which ε may be divided (3.12). Assuming, as before, a Maxwell–Boltzmann distribution law, the fraction of all collisions in which the first component of momentum along the line of centres lies between p_A and $p_A + dp_A$ and simultaneously the second component of momentum along line of centres lies between p_B and $p_B + dp_B$ is;

$$\frac{e^{-p^2/2\mu kT} \, dp_A \, dp_B}{\iint_{-\infty}^{+\infty} e^{-p^2/2\mu kT} \, dp_A \, dp_B} = \frac{e^{-p^2/2\mu kT} \, dp_A \, dp_B}{2\pi\mu kT} \tag{3.13}$$

To sum this over all distributions of p_A and p_B subject to the condition that $p^2 = p_A^2 + p_B^2$ is a constant, we see from Figure 3.3 that,

$$\int_{p^2 = p_A^2 + p_B^2} dp_A \, dp_B = \text{area of the annular strip}$$

$$= 2\pi p \, dp \tag{3.14}$$

Hence from (3.13) the fraction of collisions with a relative kinetic energy along line of centres between ε and $\varepsilon + d\varepsilon$ is,

$$\frac{dN(\varepsilon)}{N} = \frac{e^{-p^2/2\mu kT} p \, dp}{\mu kT} \tag{3.15}$$

and since, from (3.11) $p \, dp = \mu \, d\varepsilon$,

$$\frac{dN(\varepsilon)}{N} = \frac{e^{-\varepsilon/kT} \, d\varepsilon}{kT} \tag{3.16}$$

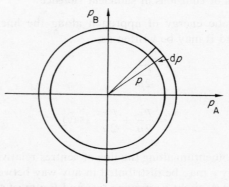

FIGURE 3.3   The calculation of $\int dp_A \, dp_B$ subject to $p^2 = p_A^2 + p_B^2 = $ constant.

The fraction of all collisions in which ε exceeds the critical value necessary for reaction, ε_c, is then,

$$\int_{\varepsilon \geqslant \varepsilon_c} \frac{dN(\varepsilon)}{N} = \int_{\varepsilon_c}^{\infty} \frac{e^{-\varepsilon/kT} \, d\varepsilon}{kT} \tag{3.17}$$

$$= e^{-\varepsilon_c/kT} \tag{3.18}$$

$$= e^{-E_c/RT} \tag{3.19}$$

where $E_c = N_{Av}\varepsilon_c$, is the critical energy per mole.

3.2c The rate of reaction

Assuming that chemical reaction does not appreciably affect the equilibrium velocity distribution the above formulae may also be used for a reacting system. For this to be true it is necessary for the rate of reaction to be much slower than the rate of collision since the Maxwell–Boltzmann distribution is maintained by the energy transfer processes that occur in the molecular collisions. Translational energy transfer between molecules takes place on every collision. As stated earlier (section 1.5) most chemical reactions are sufficiently slow because in general only a very small fraction of all collisions results in chemical reaction. Exchange of rotational and especially vibrational energy takes place much more slowly than that of translational energy and so it is not uncommon for chemical reaction to appreciably disturb the Boltzmann distribution of energy over the levels for *internal* motions of the molecules.

Combining the above results, the rate of reaction per unit volume, i.e. the rate of collisions with sufficient energy, is from (3.4) and (3.19)

$$_2r = C_A C_B Z_{AB}\, e^{-E_c/RT} \tag{3.20}$$

The second-order velocity constant $_2k$, defined by

$$_2r = _2k C_A C_B \tag{3.21}$$

is therefore given by,

$$_2k = Z_{AB}\, e^{-E_c/RT} \text{ (cc molecule}^{-1} \text{ sec}^{-1}) \tag{3.22}$$

If, as outlined above, only a fraction P of these sufficiently energetic collisions satisfy energy distribution, phase, orientation etc. requirements then

$$_2k = P Z_{AB}\, e^{-E_c/RT} \tag{3.23}$$

If it is assumed that P is independent of temperature (as is likely if it represents simply an 'orientation' factor, but not otherwise) then this result is very close to an Arrhenius rate law (1.50) since Z_{AB} varies only slowly with temperature, being proportional to $\sqrt{T}$. The experimental activation energy defined by equation (1.47) is,

$$_2E_a = -\frac{d \ln _2k}{d(1/RT)} \tag{3.24}$$

$$= -\frac{d \ln (P Z_{AB}\, e^{-E_c/RT})}{d(1/RT)}$$

$$= E_c + \tfrac{1}{2}RT \tag{3.25}$$

so that since $E_c \gg RT$ in most cases,

$$_2E_a \approx E_c \tag{3.26}$$

This relation provides some justification for the name of activation energy for E_a. The pre-exponential factor is, by (1.52),

$$_2A = _2k\, e^{+_2E_a/RT} \tag{3.27}$$

$$= P Z_{AB}\, e^{\tfrac{1}{2}} \tag{3.28}$$

$$\approx P Z_{AB} \tag{3.29}$$

Taking $P = 1$, and using typical values for collision diameters measured from transport properties the order of magnitude of A is about 10^{14} cc mole^{-1} sec^{-1}. The critical energy E_c could in principle be related to the height of the potential energy barrier separating reactant and product states, discussed in section 3.7. Since the calculation of E_c to any useful degree of approximation is not practicable for most reactions it is usually treated as an adjustable parameter when comparing the predictions of S.C.T. with experiment. As seen above (3.25) it may be set equal to the experimental activation energy to a good approximation. Comparison between theory and experiment is thus reduced to a comparison of pre-exponential factors and collision frequencies (3.28). Any discrepancy is attributed to the steric factor, P. A few simple reactions agree with S.C.T. with $P = 1$. For example the reaction

$$CH_3^{\cdot} + CH_3^{\cdot} \rightarrow C_2H_6^* \qquad [3.1]$$

which, experimentally, has zero activation energy (and would be expected to have only a very small critical energy E_c determined by the rotational energy barrier, see page 146) has a second-order rate constant $k_{obs} = 10^{13.5}$ cc mole^{-1} sec^{-1}. The calculated value of S.C.T. is $k_{calc} = 10^{13.6}$ cc mole^{-1} sec^{-1} in very good agreement. This implies that the methyl radicals in this reaction are behaving like hard spheres to a good approximation. This is the exception rather than the rule. In general most bimolecular chemical reactions require steric factors of less than unity if their rates are to be accounted for using the S.C.T. expression. For example, the abstraction reactions of hydrogen atoms

$$H + H_2 \rightarrow H_2 + H \qquad [3.2]$$

$$H + X_2 \rightarrow HX + X \qquad X = \text{halogen} \qquad [3.3]$$

$$H + RH \rightarrow H_2 + R^{\cdot} \qquad R^{\cdot} = \text{alkyl radical} \qquad [3.4]$$

all have steric factors around $0\cdot1$, which is not unreasonable for a simple orientation effect. The corresponding reactions of the more complex methyl radical, i.e.

$$CH_3^{\cdot} + H_2 \rightarrow CH_4 + H \qquad [3.5]$$

$$CH_3^{\cdot} + RH \rightarrow CH_4 + R^{\cdot} \qquad [3.6]$$

have steric factors of about $0\cdot001$. As the reactants become more complex the general trend is to lower and lower steric factors as would be expected in terms of the interpretation of P given above. Table 4.2 gives more examples.

Although most bimolecular reactions have steric factors of less than unity, some require P values greater than one to account for their observed rates. This, clearly, cannot be interpreted as being the result of additional 'steric' requirements for reaction but must result from the inapplicability of the S.C.T. expressions either for predicting the collision rate or for calculating the fraction of collisions with sufficient energy to react. The 'fast' reactions are therefore of two main types. Firstly, reactions involving charged species (e.g. ions) in which long range Coulombic forces are important and secondly, energy transfer reactions in which no net chemical change takes place but the reaction is simply the transfer of energy from one molecule to another. These will be discussed in sections 3.4 and 3.5 respectively but first the extension of S.C.T. to possible termolecular processes will be considered.

3.3 TERMOLECULAR REACTIONS

With the hard sphere model used in section 3.2 termolecular reactions are strictly impossible. Since on this picture two colliding molecules are in contact for only an infinitesimal time, the probability of a third molecule colliding with a collision pair while they are still in contact is zero. To deal with real molecules we can proceed in two ways as in (a) and (b) below.

Method (a)

We can define an elementary termolecular reaction in such a way that the hard spheres model can be used to calculate its rate. A reasonable definition of a termolecular collision is to say that a pair of molecules must be within a distance σ, the collision diameter, when a third molecule strikes one of them, i.e. the distance between the pair is insufficient to allow the third molecule to pass between them without collision (Figure 3.4). Since the average distance between collisions is λ, the mean free path, the chance of a third molecule striking the pair while they are within a distance σ apart, is σ/λ. Hence denoting the rate of termolecular collisions per unit volume by $_3r$, and the rate of bimolecular collisions per unit volume by $_2r$,

$$\frac{_3r}{_2r} = \frac{\sigma}{\lambda} \tag{3.30}$$

Now considering for simplicity a pure gas A, the mean free path is equal to the average distance travelled per second by one molecule divided by the number of collisions per second made by one molecule, i.e. from (3.2),

$$\lambda = \frac{\bar{v}_A}{\pi \sigma_A^2 \bar{v}_{AA} C_A} \tag{3.31}$$

where the symbols are as defined in section 3.2. For a Maxwell–Boltzmann velocity distribution equation (3.10) gives,

$$\frac{\bar{v}_A}{\bar{v}_{AA}} = \frac{1}{\sqrt{2}}$$

hence,

$$\lambda = \frac{1}{\sqrt{2}\pi\sigma_A^2 C_A} \tag{3.32}$$

Substituting into (3.30),

$$_3r = {}_2r \cdot \sigma_A \cdot \sqrt{2}\pi\sigma_A^2 C_A$$

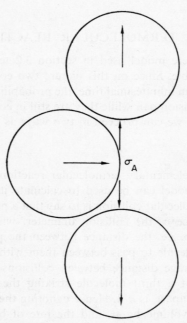

FIGURE 3.4   A 'termolecular' collision of hard spheres.

Using equations (3.4), (3.6), and (3.9) for $_2r$ gives,

$$_3r = \tfrac{1}{2}\pi\sigma_A^2 \left(\frac{8kT}{\pi^{\frac{1}{2}}m_A}\right)^{\frac{1}{2}} C_A^2 \cdot \sigma_A \cdot \sqrt{2}\pi\sigma_A^2 C_A$$

$$= \pi^2\sigma_A^5 \left(\frac{8kT}{\pi m_A}\right)^{\frac{1}{2}} C_A^3 \tag{3.33}$$

$$= Z_{AAA} C_A^3 \tag{3.34}$$

where Z_{AAA} is the termolecular collision frequency. Making the same assumptions as in 3.2 for bimolecular reactions, the third-order velocity constant, $_3k$, for a termolecular reaction is,

$$_3k = PZ_{AAA}\, e^{-E_c/RT} \qquad (3.35)$$

This assumes, in particular, that only two squared terms contribute to the critical energy, which is rather unlikely when three molecules are involved. The value of Z_{AAA} is seen from equation (3.33) to be very sensitive to the value of σ_A used, owing to the fifth power dependence, but using typical values for σ_A gives the order of magnitude of Z_{AAA} as about 10^{14} to 10^{16} cc^2 $mole^{-2}$ sec^{-1}.

Method (b)

We can consider 'termolecular' reactions to be not truly elementary but composed of two consecutive bimolecular steps. To be accurate they should then be referred to by some other name such as 'third-order reactions' since the term termolecular can only be applied to an elementary step. We write the mechanism of the reaction between A, B and C as,

$$A + B \xrightarrow{k_1} AB \qquad [3.7]$$

$$AB \xrightarrow{k_2} A + B \qquad [3.8]$$

$$AB + C \xrightarrow{k_3} products \qquad [3.9]$$

where AB is to be regarded as a short-lived, i.e. reactive intermediate, which is stable in the thermodynamic sense. Applying the steady state hypothesis to the intermediate

$$\frac{dC_{AB}}{dt} = k_1 C_A C_B - k_2 C_{AB} - k_3 C_{AB} C_C = 0$$

Therefore in the steady state

$$C_{AB} = \frac{k_1 C_A C_B}{k_2 + k_3 C_C} \qquad (3.36)$$

The rate of product formation,

$$\frac{dC_{products}}{dt} = k_3 C_{AB} C_C$$

$$= \frac{k_1 k_3 C_A C_B C_C}{k_2 + k_3 C_C} \qquad (3.37)$$

This is not a simple rate law of the form of equation (1.16) because the reaction is not elementary. However, if AB is of very low stability, i.e. it is much more likely to decompose back to A and B than it is to react with C,

$$k_2 \gg k_3 C_C \tag{3.38}$$

and we have approximately

$$\frac{dC_{products}}{dt} = \left(\frac{k_1}{k_2}\right) k_3 C_A C_B C_C \tag{3.39}$$

using (1.75)

$$= K_c k_3 C_A C_B C_C \tag{3.40}$$

where K_c is the equilibrium constant for the formation of AB. This is a simple third-order rate law with the velocity constant,

$$_3k = K_c k_3 \tag{3.41}$$

The equilibrium constant K_c may be calculated by the usual methods of thermodynamics or statistical mechanics (see Appendix A), and k_3 can be replaced by the S.C.T. expression for a bimolecular reaction, hence $_3k$ is calculated. Since K_c may take a wide range of values depending on the stability of the intermediate AB this approach is far more flexible than method (a). The experimental activation energy predicted is,

$$_3E_a = - \frac{d \ln (K_c k_3)}{d(1/RT)}$$

$$= \Delta E_p^\circ + E_3 \tag{3.42}$$

[see (1.89)] where ΔE_p° is the standard energy change in reaction [3.7] and may be negative, since the reaction is a combination process.

The most common type of third-order reaction is the combination of atoms and small radicals or molecules which need the presence of a chemically inert third body (M) to remove energy from the colliding pair of bodies to produce a stable molecule

$$A + B + M \rightarrow AB + M \tag{3.10}$$

In the absence of M the molecule AB would have sufficient energy to re-dissociate after one vibrational period. The other known third-order reactions are those involving two molecules of nitric oxide (a paramagnetic molecule which is a relatively unreactive free radical) and oxygen or a halogen,

$$2NO + O_2 \rightarrow 2NO_2 \tag{3.11}$$

$$2NO + Cl_2 \rightarrow 2NOCl \tag{3.12}$$

$$2NO + Br_2 \rightarrow 2NOBr \tag{3.13}$$

and recently the reaction

$$2I + H_2 \rightarrow 2HI \qquad [3.14]$$

has been added to the list.

Comparison of the experimental rates (see Table 4.3) with the result of method (a) shows that most atom combinations have steric factors close to unity while the reactions of NO are much slower with $P \simeq 10^{-7}$ and the hydrogen–iodine reaction [3.14] is intermediate with $P \simeq 10^{-2}$. Application of the complex mechanism (b) to the combination of atoms may involve an intermediate diatomic molecule AB* formed from the atoms A and B or alternatively the intermediate AM formed from one atom and the third body. The former is called the energy transfer mechanism for atom recombination and is identical to the reverse of the mechanism of unimolecular dissociation of the molecule AB in its second-order region which will be considered in more detail in section 3.6. The second alternative is called the atom-molecule complex mechanism. Which mechanism predominates in any particular case depends on the nature of the third body M and the strength of its chemical affinity for A. The activation energies of atom recombinations and of the nitric oxide-oxygen reaction are small and negative, i.e. the rates of these reactions decrease with increase of temperature. This can be explained by equation (3.42) since E_3 is likely to be small or zero while ΔE_p° is probably negative, for example for reactions such as

$$NO + O_2 \rightleftharpoons NO_3 \qquad [3.15]$$

However, other explanations of this unusual temperature effect are possible and will be considered later (sections 3.6 and 3.11).

3.4 REACTIONS OF CHARGED SPECIES

In S.C.T. it is assumed that the molecules of a gas do not interact except when they approach to a distance σ apart. This simple hard sphere kinetic theory of a gas works reasonably well for most normal gases as shown by the fact they follow the ideal gas laws to a good approximation at least at low pressures and high temperatures. However the presence of species carrying electrostatic charges will clearly alter this situation, since long range Coulombic forces will greatly alter the nature of collisions between the gas molecules. The electrostatic potential energy V_{es} of a pair of molecules will vary with distance apart d according to the general law

$$V_{es} = -\frac{a}{d^n} \qquad (3.43)$$

where a is a constant, depending on the charges, dipole moments, or polarizabilities, etc., which is positive for all cases except that of similarly charged ions, and n is the sum of the orders of the poles plus one, so that,

$n = 1$ for ion–ion interactions
$ = 2$ for ion–dipole interactions
$ = 3$ for dipole–dipole and ion–quadrupole interactions
$ = 4$ for ion–induced dipole (polarizable molecule) interactions
$ = 5$ for dipole–induced dipole interactions

and so on.

For example, considering the much studied case of ion–molecule reactions

$$V_{\text{es}} = -\frac{\alpha e^2}{2d^4} \qquad (3.44)$$

where e = charge on ion
$ \alpha$ = polarizability of molecule

The effect of these attractive forces on the course of collisions can be depicted as in Figure 3.5 by considering motion relative to one reactant,

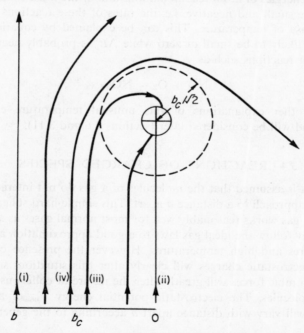

FIGURE 3.5   Some trajectories of collisions of molecules with an ion (Gioumousis and Stevenson, *J. Chem. Phys.*, **29**, 296 (1958)).

say the ion, and drawing the trajectories of the molecule for various values of the impact parameter b for fixed relative velocity. This impact parameter is the distance of closest approach of the centres of the two bodies that would occur if there were no interaction between them. At large values of b, (i), the molecule is virtually undeviated and no 'close collision' occurs. At very small values of b, (ii), the molecule undergoes a collision with the ion as in the hard sphere case. At moderate values of b the molecule either is 'captured', (iii), by the electrostatic field of the ion and spirals in to undergo close collision or else escapes after suffering appreciable deflexion, (iv). The critical value of the impact parameter, b_c, separating 'close' collisions from the rest can clearly be calculated as a function of the relative velocity, e.g. by using classical mechanics and the known potential energy function (3.44) and following the usual methods of calculating orbits in a central-force field. If the impact parameter just exceeds b_c it may be shown that the molecule and ion approach no closer than a distance $b_c/\sqrt{2}$. This minimum distance of approach and hence b_c is determined by the fact that for this critical orbit the attractive ion–molecule force just exactly balances the centrifugal force for the motion in a circle of radius $b_c/\sqrt{2}$. Equating these two forces leads to the result,

$$b_c^2 = \left(\frac{4\alpha e^2}{3\mu v^2}\right)^{\frac{1}{2}} \qquad (3.45)$$

where v is the initial relative velocity at large separation and μ is the reduced mass as before. Assuming the ion is spherically symmetric the effective collision cross-section can be obtained by rotating Figure 3.5 about the line, $b = 0$, and is clearly πb_c^2. This is the effective capturing area presented to the approaching molecules by the ion. Clearly this is smaller, the faster the molecule moves as shown by equation (3.45) above. If the number of molecules per unit volume with relative velocities between v and $v + dv$ is $dC(v)$, then the rate of collision per unit volume for these molecules with a single ion is from (3.2)

$$r(v) = \pi b_c^2 v \, dC(v) \qquad (3.46)$$

Hence the total rate of collision per unit volume (for all relative velocities) is, assuming a Maxwell distribution of velocities,

$$_2r = \frac{\int_0^\infty \pi b_c^2 v \, e^{-\frac{1}{2}\mu v^2/kT} \, dv}{\int_0^\infty e^{-\frac{1}{2}\mu v^2/kT} \, dv} \cdot C_{\mathrm{I}} C_{\mathrm{M}} \qquad (3.47)$$

$$= Z_{\mathrm{IM}} C_{\mathrm{I}} C_{\mathrm{M}} \qquad (3.48)$$

where C_I is the concentration of ions and C_M is the concentration of molecules. If, as is often the case experimentally, the velocity distribution is non-Maxwellian[‡] then the appropriate distribution function must be used instead when summing over v. By substituting equation (3.45) for b_c as a function of v into this expression and integrating, the collision frequency Z_{IM} is obtained. Many ion–molecule reactions have $E_c = 0$, and so the second-order rate constant,

$$_2k = Z_{IM} \tag{3.49}$$

For example, the reaction

$$A^+ + H_2 \rightarrow AH^+ + H \tag{3.16}$$

has an observed velocity constant of $10 \cdot 1 \times 10^{14}$ cc mole^{-1} sec^{-1}. The rate constant calculated from equation (3.49) is, $9 \cdot 0 \times 10^{14}$ cc mole^{-1} sec^{-1}, in reasonably good agreement. Because the critical impact parameter may be considerably greater than typical hard-sphere diameters the collision frequency Z_{IM} may exceed the S.C.T. value (Z_{AB}) by up to a thousand times.

3.5 ENERGY TRANSFER—THE EXTENSION OF S.C.T. FOR $2n$ SQUARED TERMS

The other main type of reactions which proceed faster than S.C.T. predicts, i.e. have $P > 1$, is the energy transfer reaction,

$$A + M \rightarrow A^* + M \tag{3.17}$$

where no chemical change occurs, but the internal energy of the reactant A is increased from a value less than ε_c to a value greater than ε_c by collision with an inert molecule M. The methods by which the rates of such processes can be studied chemically will be outlined in section 3.6 when unimolecular reactions are discussed. Clearly in this case the condition for successful 'reaction' is simply that the *total* internal energy of A be raised above the value ε_c and so one would expect more than two degrees of freedom to contribute if A is polyatomic. The fraction of collisions in which the energy in any given number of square terms exceeds ε_c must therefore be calculated to replace the fraction for two squared terms only, used in S.C.T.

Assuming that the total number of squared terms that can contribute to providing ε_c is even for simplicity, say $2n$, then the total energy can be distributed in any way over the n pairs of square terms. These n pairs may

[‡] e.g. The ion velocities may be influenced by external electric and magnetic fields.

be thought of as n oscillators each having two squared terms, one for potential and one for kinetic energy. Thus, the total energy,

$$\varepsilon = \varepsilon_1 + \varepsilon_2 + \cdots \varepsilon_n \tag{3.50}$$

where ε_1 is the energy in the first pair of squared terms etc. Equation (3.16) shows that the fraction of species (whether colliding pairs or single molecules) with energy in a single pair of squared terms between ε_1 and $\varepsilon_1 + d\varepsilon_1$ is,

$$\frac{dN(\varepsilon_1)}{N} = e^{-\varepsilon_1/kT} \, d\varepsilon_1/kT \tag{3.51}$$

Hence the fraction of species with, simultaneously, energies in the ranges ε_1 to $\varepsilon_1 + d\varepsilon_1$, ε_2 to $\varepsilon_2 + d\varepsilon_2 \ldots \varepsilon_n$ to $\varepsilon_n + d\varepsilon_n$ is,

$$e^{-\varepsilon_1/kT} \frac{d\varepsilon_1}{kT} \cdot e^{-\varepsilon_2/kT} \frac{d\varepsilon_2}{kT} \cdots e^{-\varepsilon_n/kT} \frac{d\varepsilon_n}{kT}$$

$$= \frac{e^{-\varepsilon/kT}}{(kT)^n} \, d\varepsilon_1 \, d\varepsilon_2 \ldots d\varepsilon_n \tag{3.52}$$

The total fraction, $f(\varepsilon) \, d\varepsilon$, with total internal energy between ε and $\varepsilon + d\varepsilon$ irrespective of its distribution is found by summing over all distributions of $\varepsilon_1, \varepsilon_2, \ldots \varepsilon_n$, subject to the condition that,

$$\sum_1^n \varepsilon_i = \varepsilon \tag{3.53}$$

Now, $\int \ldots \int d\varepsilon_1 \, d\varepsilon_2 \ldots d\varepsilon_n$ subject to,

$$\sum_1^n \varepsilon_i \leqslant \varepsilon \quad \text{and} \quad \varepsilon_i \geqslant 0 \text{ for all } i$$

is given by,

$$\int \cdots \int_{\sum_1^n \varepsilon_i \leqslant \varepsilon} d\varepsilon_1 \cdots d\varepsilon_n = \int_0^\varepsilon \int_0^{\varepsilon - \varepsilon_1} \cdots \int_0^{\varepsilon - \varepsilon_1 - \cdots - \varepsilon_{n-1}} d\varepsilon_n \ldots d\varepsilon_1 \tag{3.54}$$

Integration with respect to each variable in turn gives,

$$\int \cdots \int_{\sum_1^n \varepsilon_i \leqslant \varepsilon} d\varepsilon_1 \cdots d\varepsilon_n = \varepsilon^n/n! \tag{3.55}$$

The integral subject to the condition,

$$\varepsilon \leqslant \sum_1^n \varepsilon_i \leqslant \varepsilon + d\varepsilon$$

is obtained by differentiating this result with respect to ε, thus,*

$$\int \cdots \cdots \int_{\varepsilon \leqslant \sum_1^n \varepsilon_i \leqslant \varepsilon + d\varepsilon} d\varepsilon_1 \cdots d\varepsilon_n = \frac{\varepsilon^{n-1} \, d\varepsilon}{(n-1)!} \tag{3.56}$$

Hence, from (3.52),

$$f(\varepsilon) \, d\varepsilon = \int \cdots \cdots \int_{\varepsilon \leqslant \sum_1^n \varepsilon_i \leqslant \varepsilon + d\varepsilon} \frac{e^{-\varepsilon/kT}}{(kT)^n} \, d\varepsilon_1 \cdots d\varepsilon_n \tag{3.57}$$

$$= \left(\frac{\varepsilon}{kT}\right)^{n-1} \frac{e^{-\varepsilon/kT}}{(n-1)!} \frac{d\varepsilon}{kT} \tag{3.58}$$

For energy transfer reactions, since the 'reaction' is simply the *energization* of A, we can say that the probability of reaction occurring on collision equals zero if ε is less than ε_c, and equals unity, or more generally P, the steric factor, if ε is greater than or equal to ε_c. As before, when equation (3.17) was derived, the total rate of reaction is found by summing over all energies, thus the rate per unit volume is

$$_2r = PZ_{AM}C_AC_M \int_{\varepsilon_c}^{\infty} \left(\frac{\varepsilon}{kT}\right)^{n-1} \frac{e^{-\varepsilon/kT}}{(n-1)!} \frac{d\varepsilon}{kT} \tag{3.59}$$

Integrating repeatedly by parts gives the second-order rate constant,

$$_2k = {}_2r/C_AC_M$$

$$= PZ_{AM} \, e^{-\varepsilon_c/kT} \left\{ \left(\frac{\varepsilon_c}{kT}\right)^{n-1} \frac{1}{(n-1)!} + \left(\frac{\varepsilon_c}{kT}\right)^{n-2} \frac{1}{(n-2)!} \cdots + 1 \right\} \tag{3.60}$$

This result, first derived by Hinshelwood in 1936, differs from the S.C.T. result, equation (3.23), by the bracketed term. For many reactions ε_c/kT ($= E_c/RT$) is much greater than unity so that to a good approximation, taking the first term of the series only,

$$_2k = PZ_{AM} \, e^{-E_c/RT} \left\{ \left(\frac{E_c}{RT}\right)^{n-1} \frac{1}{(n-1)!} \right\} \tag{3.61}$$

---

* Using a quantum formulation this result can be obtained from the classical limit of the number of ways of distributing j quanta over n oscillators as $j \to \infty$. This number is, $(j + n - 1)!/j!(n-1)!$ which for large j, $\approx j^{n-1}/(n-1)!$

In a typical example, E_c/RT, would be about 50, so that even for only moderately complex molecules the bracketed term is very much greater than unity. For example if 10 squared terms contribute, $n = 5$, and $(E_c/RT)^{n-1} \cdot 1/(n-1)! = 50^4/4! \approx 2 \cdot 5 \times 10^5$. Hence the rates of energy transfer may exceed S.C.T. values by many orders of magnitude (see section 3.6).

The experimental activation energy derived from equation (3.61) is,

$$_2E_a = -\frac{d \ln_2 k}{d(1/RT)}$$

$$= E_c + \tfrac{1}{2}RT - (n-1)RT$$

$$= E_c - (n - \tfrac{3}{2})RT \tag{3.62}$$

For complex molecules with large values of n this may be appreciably smaller than E_c.

The question now arises as to why it is not usually necessary to use the Hinshelwood expression, (3.60), when dealing with ordinary *chemical* reactions. It seems an unlikely coincidence that in all the cases investigated so far that the reduction in rate due to 'steric' requirements is sufficient to outweigh the effect of more than two squared terms contributing to the critical energy, so that the rate is less than the S.C.T. result (3.22). A more reasonable explanation is that although initially the critical energy may be distributed over many degrees of freedom of the colliding pair of molecules, in order for chemical reaction to occur this energy must accumulate in one particular vibrational motion or pair of squared terms so that, for example, in an abstraction reaction an atom is transferred from one molecule to the other. This concept is very close to that of the statistical RRKM theories of unimolecular reactions considered in section 3.6. It will be shown there [equation (3.83)] that the probability of at least an energy ε_c out of a total energy ε accumulating in one particular pair of n pairs of squared terms is,

$$\text{probability} = \left(\frac{\varepsilon - \varepsilon_c}{\varepsilon}\right)^{n-1} \quad \text{for } \varepsilon \geqslant \varepsilon_c$$

$$\tag{3.63}$$

$$= 0 \quad \text{for } \varepsilon < \varepsilon_c$$

Replacing the discontinuous probability P of Hinshelwood's theory by this continuous function (see Figure 3.8) for a chemical reaction between A and B the rate constant is given by

$$_2k = PZ_{AB} \int_{\varepsilon_c}^{\infty} \left(\frac{\varepsilon - \varepsilon_c}{\varepsilon}\right)^{n-1} \left(\frac{\varepsilon}{kT}\right)^{n-1} \frac{e^{-\varepsilon/kT}}{(n-1)!} \frac{d\varepsilon}{kT} \tag{3.64}$$

Putting, $x = (\varepsilon - \varepsilon_c)/kT$ this becomes,

$$_2k = PZ_{AB}\, e^{-\varepsilon_c/kT}\, \frac{1}{(n-1)!} \int_0^\infty x^{n-1}\, e^{-x}\, dx$$

Integrating repeatedly by parts gives,

$$_2k = PZ_{AB}\, e^{-\varepsilon_c/kT}$$

This is the same result that is predicted by S.C.T. (3.23). This is because the statistical assumptions made in deriving (3.63) are based on the existence of energy equilibrium between the various degrees of freedom. By assuming that ε_c must accumulate eventually into one vibrational degree of freedom we get the same result as if we had assumed that the energy had to be in one pair of squared terms to start with, as in S.C.T. (3.23).

3.6 UNIMOLECULAR REACTIONS

3.6a Lindemann's theory

The first successful collision theory of unimolecular reactions was proposed in 1922 by Lindemann to replace the unsatisfactory radiation hypothesis of Perrin. In this Lindemann showed how activation of molecules by bimolecular (second-order) collisions could in certain circumstances lead to first-order kinetics for the unimolecular reaction of the molecules thus activated. The essential feature of Lindemann's mechanism was that he postulated that there is a time lag before an activated molecule can react once it has received sufficient energy to do so. During this time the activated molecule may suffer further bimolecular collisions which may de-activate it. The simplest form of this mechanism may be written,

$$A + M \xrightarrow{k_a} A^* + M \qquad \text{activation} \qquad [3.18]$$

$$A^* + M \xrightarrow{k_d} A + M \qquad \text{de-activation} \qquad [3.19]$$

$$A^* \xrightarrow{k_r} \text{products} \qquad \text{unimolecular reaction} \quad [3.20]$$

A^* is a molecule of reactant with sufficient energy (correctly distributed if necessary) to undergo spontaneous unimolecular reaction to form products. M is any molecule that can transfer energy to A by collision. In the simplest theory, all A^* molecules are treated alike and it is assumed that their specific rate of reaction, k_r, is independent of the magnitude of their activation (cf. p. 102). This mechanism is the one considered in section 1.6b as an example of the use of the steady state hypothesis. If the energy possessed by A^* is very large compared to the average energy it may be treated as a 'very reactive' intermediate in the steady state sense.

$$\frac{dC_{A^*}}{dt} = k_a C_A C_M - k_{\bar{a}} C_{A^*} C_M - k_r C_{A^*}$$

$$= 0$$

$$\therefore \qquad C_{A^*} = \frac{k_a C_A C_M}{k_{\bar{a}} C_M + k_r} \qquad (3.65)$$

The rate of formation of products,

$$\frac{dC_{products}}{dt} = r = k_r C_{A^*}$$

$$= \frac{k_a k_r C_A C_M}{k_{\bar{a}} C_M + k_r} \qquad (3.66)$$

This is not a simple rate law of the form (1.16) since the overall reaction is not strictly elementary. Although it is *chemically* elementary in that only one distinct chemical process takes place, the necessary physical steps of energy transfer to and from reactant render it a complex process overall as outlined in section 1.1b. For this reason the overall reaction is often referred to as quasi-unimolecular to distinguish it from true unimolecular elementary steps such as reaction [3.20] above.

The rate expression (3.66) will approximate to a simple rate law (1.16) under certain conditions. At high pressures of gas, the concentration C_M becomes large and so the term $k_{\bar{a}} C_M$ in the denominator of (3.66) may become very much larger than k_r, i.e. the majority of A molecules that are activated by collision are de-activated similarly before they can undergo reaction. Thus the steps of activation and de-activation are essentially balanced and the concentration of activated molecules is negligibly disturbed from its equilibrium Maxwell–Boltzmann value by the reaction [3.20]. Thus if,

$$k_{\bar{a}} C_M \gg k_r$$

$$r = \left(\frac{k_a}{k_{\bar{a}}}\right) k_r C_A \qquad (3.67)$$

which by equation (1.75)

$$= K_c k_r C_A \qquad (3.68)$$

where K_c is the equilibrium constant for activation of A,

$$K_c = \left(\frac{C_{A^*}}{C_A}\right)_e \text{ (at equilibrium)} \qquad (3.69)$$

(3.67) is a simple first-order rate law and the first-order rate constant at infinite pressure defined by

$$r^\infty = {}_1k^\infty C_A \tag{3.70}$$

is given by,

$$_1k^\infty = \left(\frac{k_a}{k_{\bar{d}}}\right) k_r \tag{3.71}$$

$$= \left(\frac{C_{A^*}}{C_A}\right)_e k_r \quad \text{[from (3.68) and (3.69)]}$$

$$= k_r f \tag{3.72}$$

where f is the equilibrium fraction of A molecules that are activated (e.g. have energy $\geqslant \varepsilon_c$). At sufficiently low pressures C_M may become small enough that,

$$k_d C_M \ll k_r$$

In this case every A molecule that is activated undergoes reaction, the time between collisions being so long that the chance of a de-activating collision occurring before A* has time to undergo reaction is negligibly small. Under these conditions,

$$r = k_a C_A C_M \tag{3.73}$$

reaction [3.18] is rate determining, and the reaction is second order overall. The concentration of activated molecules $C_{A}*$ is now very much smaller than the equilibrium value given by the Maxwell–Boltzmann distribution since collisions do not occur sufficiently rapidly to maintain this distribution. Consequently the rate of reaction is also much smaller than (3.67). The second-order rate constant at very low pressures $_2k^0$ is therefore equal to k_a and so studies of unimolecular reactions at very low pressures can lead to information about the rates of the energy transfer processes discussed earlier (section 3.5). At extremely low pressures in reaction vessels of finite size, collisions with the vessel walls may become more important than bimolecular gas-phase collisions in activating the reactant molecules. Under these conditions the rate of reaction should tend to first-order kinetics, since the rate of collision with the wall is proportional to the gas pressure. For example in a vessel of about 10 cm diameter the mean free gas-phase path of a typical gas molecule exceeds the vessel diameter at pressures below about 5×10^{-4} torr. At pressures below this heterogeneous wall activation will exceed homogeneous gas-phase activation, and if activation is rate determining the first-order rate constant will be proportional to the surface area to volume ratio of the reaction vessel. Comparatively few quantitative kinetic investigations have been made of unimolecular reactions in this very low pressure range at present, but as outlined in section 2.3b there is in principle no difficulty in doing so with mass-spectrometric techniques. The effect of pressure on the homogeneous reaction has been much more extensively studied in those cases where the fall-off in rate occurs in an experimentally convenient pressure region.

All unimolecular reactions that would be expected to show this phenomenon of changing order from one to two as the pressure is reduced from high to low values, do so. Consequently, the reverse reactions also show changes in order as the pressure is varied, since as outlined in section 1.5, detailed balancing, equation (1.75), applies even to non-equilibrium systems, provided only that the translational equilibrium is undisturbed. In most experimental conditions this requirement is satisfied and so isomerization reactions,

$$A \rightarrow B \qquad\qquad [3.21]$$

change from first to second order in both directions as the pressure is reduced, while dissociation reactions for example

$$A \rightarrow B + C \qquad\qquad [3.22]$$

have reverse processes (combinations) which change from second to third order as the pressure is lowered (see page 97).

Since M may be any species, whether reactant, product, or added inert gas the change in rate of reaction and reaction order of a unimolecular reaction or its reverse can be brought about by adding or removing *any* gas. For example if the pressure of a pure reactant is such that its unimolecular reaction is not in the high pressure first-order region (i.e. it is in what is called the fall-off region) and a large amount of a chemically inert gas is added, this will result in the rate of reaction increasing towards the high pressure limiting value, $_1k^\infty C_A$, as the concentration of activated molecules is increased towards its equilibrium value. The efficiency of a gas in increasing the rate towards the limit will, of course, vary from gas to gas, since the values of k_a and $k_{\bar{a}}$ depend on the nature of M. In this way the relative efficiencies of molecules as energy transfer agents may be investigated. This effect of a chemically inert substance on the rate of a chemical reaction is an important diagnostic test for unimolecular reactions, although other complex systems may show similar effects (e.g. in radical chain reactions adding an inert gas may reduce the rate of diffusion of active centres to the vessel wall and hence accelerate the reaction). Since the energy transfer efficiencies of the products of a unimolecular reaction are often not very different from that of the reactant the variation of reaction rate with time in a closed constant-volume system follows quite accurately a first-order law, even to quite high percentage conversions, over the whole range of reactant pressures. That is, the time-order remains unity even when fall-off has occurred and the true order is approaching two. This may be seen from equation (3.73) if C_M is approximately independent of time, see p. 11.

The region of the pressure scale at which these changes of order occur can be interpreted in terms of the rate constants of the elementary processes as follows. Define the first-order constant, $_1k$, as,

$$_1k = r/C_A \quad \text{(at all pressures)}$$

then, from (3.66),

$$_1\text{k} = \frac{k_a k_r C_M}{k_d C_M + k_r} \tag{3.74}$$

(3.72) gives,

$$= \frac{k_r f}{1 + k_r/k_d C_M} \tag{3.75}$$

The pressure at which $_1k$ has fallen to one half of its high pressure limiting value, $_1k^\infty$, given by (3.72), is determined by the condition that,

$$k_d {}_\frac{1}{2}C_M = k_r \tag{3.76}$$

where $_\frac{1}{2}C_M$ is the concentration of M at which this occurs, hence

$$_\frac{1}{2}C_M = k_r/k_d$$

(3.72) gives,

$$= {}_1k^\infty/k_a \tag{3.77}$$

Measurement of $_\frac{1}{2}C_M$ and $_1k^\infty$ therefore allows k_a to be calculated even in cases where results at sufficiently low pressures are not available to allow calculation of k_a from the second-order limiting low pressure constant $_2k^0$. As outlined in section 3.5 to interpret the values thus measured for k_a requires that more than two squared terms be allowed to contribute to activation, for example as in Hinshelwood's equation (3.60). If the S.C.T. expression for two squared terms, equation (3.23), were used for k_a in order to predict $_\frac{1}{2}C_M$ from measurements of $_1k^\infty$, the resulting pressure would in general be many orders of magnitude higher than that observed experimentally. It was this fact which led Hinshelwood to formulate the theory outlined in 3.5.

Although this simple Lindemann theory can thus explain many of the important general characteristics of unimolecular reactions it fails when more detailed quantitative aspects of these reactions are investigated. The most serious discrepancy occurs when the actual manner in which the reaction rate changes from first to second order is considered. The first-order constant $_1k$ is given by Lindemann's theory, [equation (3.74)] as

$$_1k = \frac{k_a k_r C_M}{k_d C_M + k_r}$$

$$\therefore \qquad 1/_1k = \frac{k_d}{k_a k_r} + \left(\frac{1}{k_a}\right)\frac{1}{C_M} \qquad (3.78)$$

or from (3.71),

$$\frac{_1k^\infty}{_1k} = 1 + \left(\frac{k_r}{k_d}\right) \cdot \frac{1}{C_M} \qquad (3.79)$$

Thus a plot of $(_1k)^{-1}$, or $(_1k^\infty/_1k)$, versus the reciprocal of pressure should be linear. In fact as can be seen for example in Figure 3.6 the experimental

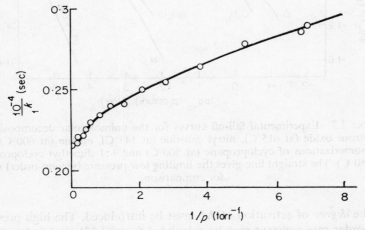

FIGURE 3.6   A plot of $1/_1k$ versus $1/p$ for 1:1 dimethyl cyclopropane isomerization (Flowers and Frey, *J. Chem. Soc.*, 1160 (1962)).

plots are distinctly curved. The fall-off in rate in practice is much slower than expected from the simple equation (3.74). Thus a change of pressure by a factor of 100 should change the relative magnitudes of the two terms in the denominator of (3.74) from

$$k_d C_M = 10 \times k_r \text{ (say)}$$

to, $$k_d C_M = \tfrac{1}{10} \times k_r$$

i.e. from very close to first-order kinetics to very close to second-order kinetics. In fact for complex molecules the fall-off in rate from first to second-order takes place over many more than two orders of magnitude of pressure, as can be seen from illustrative fall-off curves in Figure 3.7. In general the more complex the molecule the more pronounced this discrepancy becomes. To account for this observation a more detailed theory than the simple Lindemann mechanism which considers only two classifications of reactant molecules, i.e. normal and activated, must be used. Some allowance that the rate of unimolecular reaction, k_r, depends

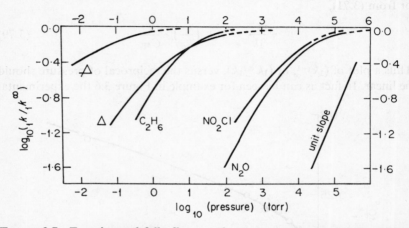

FIGURE 3.7 Experimental fall-off curves for the unimolecular decompositions of nitrous oxide (at 615°C), nitryl chloride (at 147°C), ethane (at 600°C) and the isomerizations of cyclopropane (at 500°C) and 1:1 dimethyl cyclopropane (at 480°C). The straight line gives the limiting low pressure (second-order) slope for comparison.

on the *degree* of activation of A* must be introduced. The high pressure first-order rate constant may be calculated from (3.72) modified to allow for this dependence, thus,

$$_1k^\infty = \Sigma\, k_r f \tag{3.80}$$

where f is the equilibrium fraction of A in a specified activated state, and k_r is the rate constant for reaction of that state and the sum is taken over all activated states of A, i.e. states for which $k_r \neq 0$. Similarly from (3.75) the general first-order constant is,

$$_1k = \Sigma\, \frac{k_r f}{1 + k_r/k_d C_M} \tag{3.81}$$

In expressions (3.80) and (3.81), f is readily calculated statistically since it is an equilibrium property. The rate of deactivation, k_d, is usually taken to be equal to the collision frequency Z_{AM} since it is often assumed that deactivation occurs on every collision of activated molecules. This is a consequence of the very high energies associated with activation for most reactions when compared to the average thermal energies of non-activated molecules. Such high energies are much more likely to be lost during a collision with a molecule of low energy than they are to be enhanced. Although multi-step deactivation (and, therefore, also, activation) has been considered, the theories that are discussed now take the simpler view that $k_d = Z_{AM}$. The problem, therefore, reduces to one of calculating k_r as a function of the state of activation. Some attempts to do this calculation are outlined below.

3.6b Statistical, Rice–Ramsperger–Kassel–Marcus (RRKM) theory.

We shall only consider the simplest version of this theory to illustrate the type of physical model used. The molecule of reactant is regarded as a system of n_k loosely coupled degenerate simple harmonic oscillators of vibration frequency v. The loose coupling of the oscillators allows free flow of energy between them to occur with a frequency v, without introducing significant anharmonicity. Reaction is assumed to occur when at least an energy ε_c out of the total energy of the oscillators, ε, by chance finds its way into one particular oscillator. Clearly if $\varepsilon < \varepsilon_c$ the chance of this happening is zero, but if $\varepsilon \geqslant \varepsilon_c$, there is a definite probability of this accumulation occurring. The theory is a statistical one in that all *detailed* arrangements of energy between the oscillators are considered equally likely provided the total energy is constant. This fundamental assumption of statistical mechanics is certainly justified by results (i.e. indirectly) for equilibrium systems. It is extended to kinetic systems by assuming that when energy is transferred between the oscillators it is equally likely to produce any one of the various detailed arrangements corresponding to a given total energy. Thus the probability of reaction occurring when a transfer of energy between the oscillators takes place is equal to the number of detailed ways of arranging the energy with *at least* an energy ε_c in one particular oscillator (pair of squared terms) divided by the total number of possible arrangements (i.e. with *any* amount of energy, less than the total ε, in the selected oscillator). Using the high temperature classical statistics of section 3.5, the fraction of molecules with a total energy $\sum_{1}^{n_k} \varepsilon_i$ given by

$$\sum_{1}^{n_k} \varepsilon_i \leqslant \varepsilon$$

is evaluated with the aid of equation (3.55)

$$\int \cdots \int\limits_{\sum\limits_{1}^{n_k} \varepsilon_i \leqslant \varepsilon} d\varepsilon_1 \, d\varepsilon_2 \cdots d\varepsilon_{n_k} = \frac{\varepsilon^{n_k}}{n_k!}$$

Hence the probability of reaction equals the number of ways of arranging the energy with at least ε_c fixed in the first oscillator divided by the number of ways of arranging the energy with any amount, less than ε, in the first oscillator. This is the same as the number of ways of distributing the energy of $(\varepsilon - \varepsilon_c)$ or less over the remaining $(n_k - 1)$ oscillators, divided by the number of ways of distributing an energy of ε or less over $(n_k - 1)$ oscillators. Thus the probability of reaction is given by

$$\frac{\displaystyle\int \cdots \int\limits_{\sum\limits_{2}^{n_k} \varepsilon_i \leqslant \varepsilon - \varepsilon_c} d\varepsilon_2 \, d\varepsilon_3 \cdots d\varepsilon_{n_k}}{\displaystyle\int \cdots \int\limits_{\sum\limits_{2}^{n_k} \varepsilon_i \leqslant \varepsilon} d\varepsilon_2 \, d\varepsilon_3 \cdots d\varepsilon_n} \qquad (3.82)$$

(3.55) gives,   probability $= \dfrac{(\varepsilon - \varepsilon_c)^{n_k-1}/(n_k - 1)!}{\varepsilon^{n_k-1}/(n_k - 1)!}$

$$= \left(\frac{\varepsilon - \varepsilon_c}{\varepsilon}\right)^{n_k-1} \qquad (3.83)$$

This reaction probability as a function of energy ε is plotted in Figure 3.8 for several values of n_k. In the case $n_k = 1$, (two squared terms) it corresponds to the simple step function assumed in the Lindemann–Hinshelwood treatment (section 3.6a). For $n_k > 1$, the distinction between activated and non-activated molecules becomes less clear-cut. For very complex

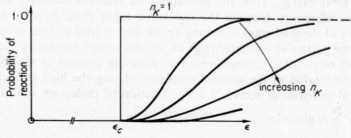

FIGURE 3.8   Plot of the probability $\left(\dfrac{\varepsilon - \varepsilon_c}{\varepsilon}\right)^{n_k-1}$ versus ε for various n_k.

molecules (large n_k) the reaction probability increases slowly with energy and remains quite small even for energies appreciably above the critical energy ε_c. Now since energy is reshuffled ν times per second the specific reaction rate k_r is equal to ν times the probability that a reshuffle causes reaction, i.e.

$$k_r = \nu \left(\frac{\varepsilon - \varepsilon_c}{\varepsilon} \right)^{n_k - 1} \tag{3.84}$$

This rate constant is a function of total energy, ε, only and does not depend on the distribution of the energy over the oscillators or on such factors as the vibrational phases of the oscillators. This is a consequence of the statistical assumption outlined above that total energy is the only relevant parameter in determining the relative probability of formation of the various detailed arrangements. Justification of this assumption in non-equilibrium systems is impossible except by results.

The fraction of molecules with total energy between ε and $\varepsilon + d\varepsilon$ in $2n_k$ squared terms is by equation (3.58)

$$f(\varepsilon)\, d\varepsilon = \left(\frac{\varepsilon}{kT} \right)^{n_k - 1} \frac{e^{-\varepsilon/kT}}{(n_k - 1)!} \frac{d\varepsilon}{kT}$$

And hence from (3.80) and (3.84), replacing the sum by the integral,

$$_1 k^\infty = \int_{\varepsilon_c}^\infty k_r f(\varepsilon)\, d\varepsilon \tag{3.85}$$

$$= \int_{\varepsilon_c}^\infty \nu \left(\frac{\varepsilon - \varepsilon_c}{\varepsilon} \right)^{n_k - 1} \left(\frac{\varepsilon}{kT} \right)^{n_k - 1} \frac{e^{-\varepsilon/kT}}{(n - 1)!} \frac{d\varepsilon}{kT}$$

evaluating the integral as before [(3.64)] gives

$$_1 k^\infty = \nu\, e^{-\varepsilon_c/kT} \tag{3.86}$$

This result for the high pressure first-order rate constant is exactly of the Arrhenius form (1.50) with,

$$_1 A^\infty = \nu$$
$$_1 E_a^\infty = \varepsilon_c N_{\mathrm{Av}} \tag{3.87}$$
$$= E_c \text{ per mole}$$

both parameters being independent of temperature. Molecular vibration frequencies are typically of the order of magnitude $10^{14 \pm 1}$ sec^{-1}, so that pre-exponential factors for first-order unimolecular reactions are expected to be in this range. Table 4.1 gives some examples. One reason for choosing the system of units cc, moles and seconds is that 'normal' pre-exponential factors for unimolecular, bimolecular and termolecular reactions then all have the same magnitude (10^{14}) within a few powers of ten.

In the general pressure case equation (3.81) becomes.

$$_1k = \int_{\varepsilon_c}^{\infty} \frac{k_r f(\varepsilon)\, \mathrm{d}\varepsilon}{1 + k_r/k_d C_M} \tag{3.88}$$

Assuming deactivation on every collision, so that $k_d = Z_{AM}$, from (3.58) and (3.84)

$$_1k = \int_{\varepsilon_c}^{\infty} \nu \frac{\left(\dfrac{\varepsilon - \varepsilon_c}{\varepsilon}\right)^{-1}\left(\dfrac{\varepsilon}{kT}\right)^{n_k-1} \dfrac{\mathrm{e}^{-\varepsilon/kT}}{(n_k - 1)!} \dfrac{\mathrm{d}\varepsilon}{kT}}{1 + \nu\left(\dfrac{\varepsilon - \varepsilon_c}{\varepsilon}\right)^{n_k-1} \Big/ Z_{AM} C_M}$$

which simplifies to,

$$_1k = \frac{\nu\, \mathrm{e}^{-\varepsilon_c/kT}}{(n_k - 1)!} \int_0^{\infty} \frac{x^{n_k-1} \mathrm{e}^{-x}\, \mathrm{d}x}{1 + \dfrac{\nu}{Z_{AM} C_M}\left(\dfrac{x}{b + x}\right)^{n_k-1}} \tag{3.89}$$

where, $x = (\varepsilon - \varepsilon_c)/kT$

and $b = \varepsilon_c/kT$

or

$$\frac{_1k}{_1k^{\infty}} = \frac{1}{(n_k - 1)!} \int_0^{\infty} \frac{x^{n_k-1} \mathrm{e}^{-x}\, \mathrm{d}x}{1 + \dfrac{\nu}{Z_{AM} C_M}\left(\dfrac{x}{b + x}\right)^{n_k-1}} \tag{3.90}$$

This integral may be evaluated numerically and hence a plot of log $(_1k/_1k^{\infty})$ versus log (pressure) constructed as in Figure 3.9. This formula

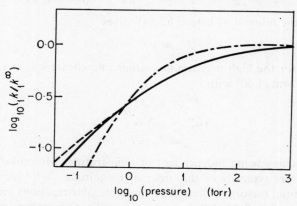

FIGURE 3.9   A comparison of the fall-off curves predicted for the isomerization of cyclopropane at 500°C by the, Hinshelwood–Lindemann theory------; Slater theory ——; Kassel theory----. (Pritchard, Sowden and Trotman-Dickenson, *Proc. Roy. Soc.*, A217, 567 (1953)).

gives excellent agreement with most experimental data provided n_k is treated as an adjustable parameter and chosen to give the best fit. If this is done the value for n_k obtained is generally about one half the maximum possible value, which is the number of normal modes in the molecule.

At low pressures $Z_{AM}C_M$ in equation (3.89) tends to zero so that,

$$_1k^0 = \frac{\nu\,e^{-\varepsilon_c/kT}}{(n_k-1)!}\frac{Z_{AM}C_M}{\nu}\int_0^\infty (b+x)^{n_k-1}e^{-x}\,dx \qquad (3.91)$$

or for the second-order constant at low pressure

$$_2k^0 = {}_1k^0/C_M$$

$$= Z_{AM}\,e^{-\varepsilon_c/kT}\left\{\left(\frac{\varepsilon_c}{kT}\right)^{n_k-1}\frac{1}{(n_k-1)!}+\cdots+1\right\} \qquad (3.92)$$

This is the same result deduced earlier, (3.60), from Hinshelwood's theory. As before, the low pressure activation energy is,

$$_2E_a^0 = -\frac{d\ln {}_2k^0}{d(1/RT)}$$

$$= \frac{E_c + \tfrac{1}{2}RT}{\left\{1+(n_k-1)\dfrac{RT}{E_c}+\cdots(n_k-1)!\left(\dfrac{RT}{E_c}\right)^{n_k-1}\right\}} \qquad (3.93)$$

which if $\dfrac{(n_k-1)RT}{E_c}$ is small, approximates to

$$_2E_a^0 = E_c - (n_k - \tfrac{3}{2})RT \qquad (3.94)$$

as before. Thus the activation energy decreases from its high pressure value, E_c, by an amount determined by n_k. If the number of effective oscillators is large this decrease is measurable. For example Table 3.1 shows some

TABLE 3.1 The variation of activation energy of the isomerization of cyclopropane with pressure

Pressure torr	∞	50	10	5
E_a kcal mole^{-1}	$65\cdot60 \pm 0\cdot03$	$64\cdot17 \pm 0\cdot15$	$61\cdot80 \pm 0\cdot35$	$60\cdot79 \pm 0\cdot37$

results obtained for the isomerization of cyclopropane to propylene. This decrease corresponds to a value for n_k of about 7. Considering both forward and backwards reactions equation (1.91) gives

$$\Delta E^\circ = E_{af} - E_{ab}$$

$$= \text{a constant, independent of pressure}$$

Hence, if E_{af} decreases with decreasing pressure, then so must E_{ab}. Sometimes e.g. for atom and radical recombinations (reversed dissociations), $E_{cb} = 0$ and then E_{ab} at low pressures becomes negative thus,

$$_3E_{ab}^0 = E_{cb} - (n_k - \tfrac{3}{2})RT$$
$$= -(n_k - \tfrac{3}{2})RT$$

This is a way of explaining negative activation energies for such reactions from a different view-point to that given in section 3.3. A third possible explanation will be given later in section 3.11.

A more general formulation, than the simple Kassel theory given here, leads to the equation corresponding to (3.84)

$$k_r \propto \frac{w(\varepsilon - \varepsilon_c)}{w(\varepsilon)} \tag{3.95}$$

where $w(\varepsilon)$ is the density of states (statistical weight factor, see Appendix A) for a non-fixed energy ε. Marcus used this expression in formulating his theory. Quantum versions of this theory have been given and it has also been extended to non-degenerate oscillators. However, the simple picture given above illustrates the basic idea behind all these theories. We now consider a rather different approach to the problem formulated by Slater.

3.6c Slater's theory

In contrast to the RRKM theory this is a dynamical theory in which the rate of decomposition of activated molecules is calculated using classical mechanics, without the 'statistical' assumption (p. 111). The model of a molecule of reactant that is used in this theory is one consisting of a number of classical, simple harmonic (normal mode) oscillators of various frequencies. These normal modes are taken to be strictly harmonic so that no energy transfer can occur between them. Thus it is assumed that the amount of energy in each normal mode stays fixed between collisions and can only change when collision occurs. Decomposition or isomerization is assumed to occur when the phases of the vibrations are such that one particular internal coordinate (q) of the molecule exceeds some critical value (q_c). This coordinate may be the distance between two atoms, a bond angle or any linear combination of any number of these coordinates. This coordinate must be chosen with reference to the motion of the atoms that is believed to accompany the reaction and clearly in most cases some uncertainty exists in making this choice. Since after an activating collision has occurred no energy flow between the normal modes is possible in this theory, not only must the activated molecule possess an energy at least equal to ε_c but this energy must be correctly distributed at the instant of activation if this molecule is to undergo subsequent reaction. This activation requirement is clearly more stringent than that of RRKM theory and so rates of activation in the Slater sense must be lower. Also the rate of reaction k_r is no longer simply a function of total energy only, but it depends on the way the energy is distributed over the normal modes. The critical coordinate will be a linear combination of a number, say n_s, of simple harmonic normal modes, of different frequencies, and therefore, constantly varying relative phases. A plot of this coordinate q against time since the last collision might look somewhat like Figure 3.10. If the initial phases of

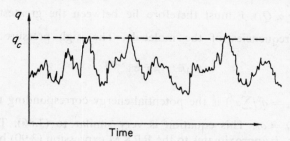

FIGURE 3.10   A plot of the sum of a large number of simple harmonic motions of different frequencies, such as might correspond to a typical coordinate q.

motion are randomly distributed the phase-averaged time lag after collision before the critical value of the coordinate q_c is reached in an upward going direction can be calculated since this is the same as the average time interval between successive upward crossings of the line $q = q_c$ for a non-reacting (truly simple harmonic) molecule. If the distribution of phases immediately after collision is not uniformly random or if the crossing points are not randomly spaced, but occur say in groups, the calculation of the average time lag becomes more difficult. These statistical assumptions simplify the calculation but even so the mathematics involved is complex and will not be given here.

The results of this theory when made the subjects of certain approximations are rather similar to those of RRKM theory, but with certain important differences. Firstly, the number of normal modes n_s is the number that may contribute to altering the coordinate q, i.e. it includes all of the normal modes of the molecule except those which for symmetry reasons cannot affect q. Secondly, the formula for k_r as a function of total energy becomes,

$$\bar{k}_r = \bar{\nu} \left(\frac{\varepsilon - \varepsilon_c}{\varepsilon} \right)^{n_s - 1} \qquad (3.96)$$

only after it has been averaged over all distributions of the internal energy. The frequency $\bar{\nu}$ is given by

$$\bar{\nu}^2 = \frac{\sum_1^{n_s} \alpha_i^2 \nu_i^2}{\sum_1^{n_s} \alpha_i^2} \tag{3.97}$$

It is a weighted root mean square of all the normal mode frequencies ν_i, the weighting (amplitude) factors α_i representing the magnitude of the contribution of each normal mode coordinate Q_i to the displacement q (i.e. $q = \sum_1^{n_s} \alpha_i Q_i$). $\bar{\nu}$ must therefore lie between the greatest and least vibration frequencies of the molecule. The high pressure first-order constant is

$$_1k^\infty = \bar{\nu}\, e^{-\varepsilon_c/kT} \tag{3.98}$$

where ε_c ($\varepsilon_c = q_c^2/\sum_1^{n_s} \alpha_i^2$) is the potential energy corresponding to the configuration $q = q_c$. This equation is very similar to (3.86). The general formula for $_1k$ approximates to the RRKM expression (3.90) but with the substitution,

$$n_k = \tfrac{1}{2}(n_s + 1) \tag{3.99}$$

Since n_s is usually not very different from the total number of modes, while a value of n_k of about half this maximum often gives best fit to experimental data, it can be seen from (3.99) that the predictions of the two theories about the shape of the fall-off curves do not differ greatly in many cases. This is illustrated in Figure 3.9.

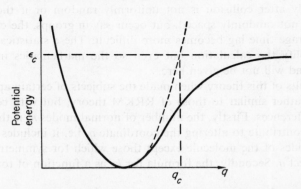

FIGURE 3.11 A comparison of the potential energy curves for a simple harmonic oscillator and a typical molecular vibrational mode. S.H.O., ----; molecule ——.

All the experimental evidence is that energy flow *does* occur freely between the normal modes of real molecules. This is because at the high energy levels associated with chemical reactions the molecular vibrations are considerably anharmonic. The parabolic potential curve is a valid approximation for the low vibrational levels only as illustrated in Figure 3.11. Hence rates of activation calculated on Slater's theory must be too low since they do not allow for this redistribution of energy between the normal modes. On the other hand the view that reaction occurs when a coordinate reaches a particular value, e.g. a bond length reaches a critical extension, is a more reasonable picture of a chemical reaction than the RRKM criterion that a certain amount of energy must accumulate in a particular oscillator. A molecule could well have more energy than ε_c in this mode and yet its configuration might not be such that it could be said to have undergone reaction (see Figure 3.11). In an attempt to reconcile these viewpoints Gill and Laidler have proposed a hybrid of the two theories.

3.6d Gill and Laidler

The mechanism envisaged is summarized as

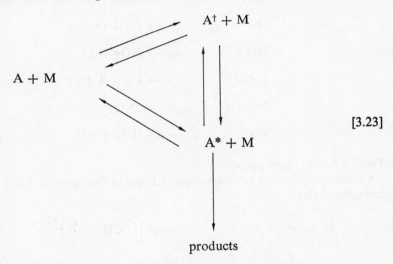

$$A^\dagger + M$$

$$A + M$$

[3.23]

$$A^* + M$$

products

Where $A^\dagger$ is a reactant molecule that is activated in the RRKM sense and A^* is one that is activated in the more stringent Slater sense. Only the latter is assumed to be able to react to give products, but an RRKM activated molecule may become an A^* molecule by internal energy transfer (without a collision). Free flow of energy is assumed in accordance with

5

experiment but the criterion for reaction is the Slater one of coordinate extension. At high pressures this mechanism leads to the same results as Slater's theory, but at low pressures the results of RRKM theory are obtained. This is because at low pressures an $A^\dagger$ molecule always has time to reorganize its energy distribution to form A^* and hence to react before it is deactivated by collision. At high pressures this is not so and only A^* molecules formed directly by collision are likely to produce products. Since the differences between the predictions of Slater and RRKM theories are not easily detected experimentally (see Figure 3.9 for example), both can be fitted to many results within experimental error. It follows that this combined theory also can give good agreement with experiment.

Some examples of reactions of stable molecules which have been found to fit in well with these unimolecular rate theories are:

$$\text{cyclopropane} \longrightarrow \text{propylene} \qquad [3.24]$$

$$\text{cyclobutane} \longrightarrow 2C_2H_4 \qquad [3.25]$$

$$N_2O \longrightarrow N_2 + O \qquad [3.26]$$

$$C_2H_6 \longrightarrow 2CH_3^\cdot \qquad [3.27]$$

$$O_3 \longrightarrow O_2 + O \qquad [3.28]$$

$$NO_2Cl \longrightarrow NO_2 + Cl \qquad [3.29]$$

$$C_2H_5Cl \longrightarrow C_2H_4 + HCl \qquad [3.30]$$

$$N_2O_4 \longrightarrow 2NO_2 \qquad [3.31]$$

$$N_2O_5 \longrightarrow NO_2 + NO_3 \qquad [3.32]$$

Table 4.1 gives more examples.

Chemical activation of unimolecular reaction (see section 2.2d) using reactions such as,

$$H + {>}C = C{<} \longrightarrow \left({>}CH{-}C{\overset{\cdot}{<}} \right)^* \qquad [3.33]$$

$$D + {>}C = C{<} \longrightarrow \left({>}CD{-}C{\overset{\cdot}{<}} \right)^* \qquad [3.34]$$

or

$$CH_2 + {>}C = C{<} \longrightarrow \left(\overset{H \quad H}{\underset{\triangle}{\diagup\diagdown}} \right)^* \qquad [3.35]$$

to produce activated molecules, can yield information about the ease of energy flow between normal modes. It can also be used to measure the variation of k_r with ε and hence to check equations such as (3.84) and (3.96).

3.7 OTHER COLLISION THEORIES OF BIMOLECULAR REACTIONS

In this section a brief outline of attempts to formulate more detailed collision theories of reaction rates than S.C.T. is given. In particular it is desirable to deal with reactions in systems with non-equilibrium energy distributions. Transition state theories, which are described in sections 3.8 to 3.13 provide a more detailed interpretation of reaction rates in equilibrium systems than is given by S.C.T. but cannot be applied under non-equilibrium conditions. This has caused revived interest in collision theory during the last ten years or so. Many systems of great experimental importance have non-Boltzmann energy distributions. For example, reactions in shock waves, flames, and crossed molecular beams, as well as ion-molecule reactions in electric fields which occur in mass-spectrometers all take place under non-equilibrium conditions.

Owing to the complexity of molecular collisions we will do little more here than define the problem that is to be solved and outline one way in which partial solution has been achieved.

3.7a Collision cross-sections

Figure 3.12 illustrates diagramatically the steps that must be taken when a rate constant is calculated by general collision theory. First we must calculate P_{ij}^{kl} (E, b, β, θ, ϕ, K), the probability that reaction will occur when the reactants I and J in internal energy states represented by i, and j, collide with an initial relative kinetic energy E, and impact parameter b, in a plane defined by β to form products K + L in internal energy states k and l which are scattered in a direction defined by θ and ϕ with respect to the direction of the initial kinetic energy, via a collision complex ('super molecule' or pair of molecules in close proximity) in energy state K. This probability must then be integrated over all states of the complex K, and all impact parameters b and β to obtain I_{ij}^{kl} (E, θ, ϕ) which is called the *differential reaction cross-section* by analogy with the hard sphere case. This when integrated over all scattering angles of the products gives $C_{ij}^{kl}(E)$ the *total reaction cross-section*. Integration of this over the appropriate translational energy distribution then gives $k_{ij}^{kl}(T)$ the *detailed specific rate constant* (as a function of temperature if translational energy is equilibrated). A weighted average of this over all available energy states leads to the ordinary specific rate constant $k(T)$. Before this sequence of operations can be carried out the first step, the calculation of reaction

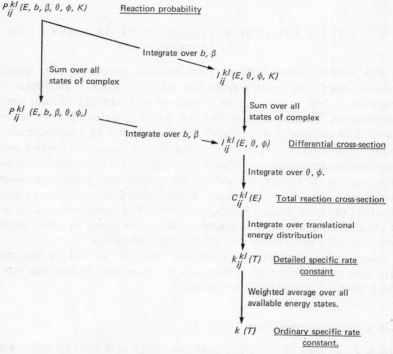

FIGURE 3.12 The relations between rate constants and reaction probabilities on general collision theory (Eliason and Hirschfelder, *J. Chem. Phys.*, **30**, 1426 (1959)).

probability P from the dynamics of the collision, must be performed. One method of achieving this aim using the concept of potential energy surfaces will be described.

3.7b Potential energy surfaces

Owing to the large difference in mass of the electron and atomic nuclei the wave equation describing a molecule or pair of interacting molecules may be simplified by the Born–Oppenheimer approximation which separates the electronic from the nuclear wave equations. Thus the stationary wave equation ($\mathcal{H}\Psi = W\Psi$) may (in principle) be solved for given fixed positions of the atomic nuclei in the molecule to obtain the electronic energy states W_i. Addition to these of the internuclear repulsion energy† yields the molecular potential energy U_i. Repetition of this calculation for all possible nuclear positions thus leads to U_i as a function of all the internal coordinates of the molecule. For example, a diatomic molecule has only one internal coordinate (the distance between the atomic nuclei, d) and a plot of U_i versus d is called the potential energy curve of the molecule in the electronic state i. A typical ground state ($i = 0$) potential energy curve for a diatomic molecule is similar to the solid curve of Figure 3.11. For a polyatomic molecule of N atoms there are $(3N - 6)$ internal coordinates if the molecule is non-linear, since 3 coordinates must be specified to define the position of each nucleus of the molecule and of these $3N$ coordinates 3 define the position of the molecule as a whole in space (e.g. by the Cartesian coordinates of the centre of gravity) and 3 define the orientation of the molecule in space (e.g. by the 3 Eulerian angles relative to chosen axes). For linear molecules there is one extra internal coordinate, i.e. $(3N - 5)$ since only two coordinates are needed to specify the orientation of the molecule as a whole. For polyatomic molecules the variation of U_i with internal coordinates can, therefore, be represented by a *potential energy hypersurface* drawn in $1 + (3N - 6)$ [or for a linear molecule $1 + (3N - 5)$] dimensions. If there is no molecular rotation the potential energy surface thus defined governs the internal motions (vibrations) of the nuclei of the molecule, and these motions can be treated as those of particles in a potential field given by U_i.

The effects of rotational motion can be approximately included (ignoring Coriolis forces) by adding to U_i a centrifugal potential, to give an effective potential energy function V_i where, for example, for a diatomic molecule

$$V_i = U_i + \frac{J(J + 1)h^2}{8\pi^2 I} \tag{3.100}$$

Where J is the rotational quantum number, and $I = \mu d^2$ is the moment of inertia. This function is illustrated in Figure 3.13. The curve with the horizontal inflexion corresponds to the rotational level at which centrifugal force just causes the molecule to become unstable.

The motion of a pair of colliding atoms can be represented by the motion of a representative point mass along these potential energy curves.

---

† This is the sum of all terms such as

$$\frac{Z_A Z_B e^2}{d_{AB}}$$

for the potential energy due to repulsion between all pairs of nuclei A and B of charges $Z_A e$ and $Z_B e$ separated by distance d_{AB}.

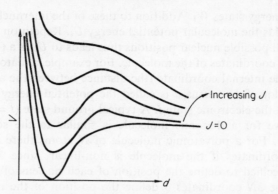

FIGURE 3.13 The effective potential energy (including rotational energy) for a diatomic molecule.

However, such a system cannot undergo chemical reaction (see section 3.3). For the simplest bimolecular chemical reaction between an atom and a diatomic molecule,

$$A + BC \rightarrow AB + C \qquad [3.36]$$

for example,

$$H + H_2 \rightarrow H_2 + H \qquad [3.37]$$

(which may be studied by observing the conversion of para to ortho hydrogen, section 4.3b) the collision complex or super molecule is triatomic and the potential energy is, therefore, a function of $3 \times 3 - 6 = 3$,

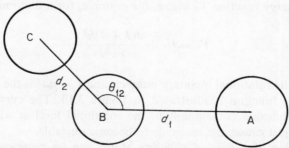

FIGURE 3.14 A possible set of internal coordinates for a triatomic collision complex.

coordinates. For example we could choose the three shown in Figure 3.14 so that $V = V(d_1, d_2, \theta_{12})$. To plot out this function one would need to use four dimensions, so for simplicity only linear systems ($\theta_{12} = \pi$) are considered for which V is a function of d_1 and d_2 only. It can be shown that for reactions such as [3.37] linear configurations have a lower energy and therefore greater probability than non-linear ones so that it is not unreasonable to choose the linear system for illustrative purposes. With this simplification the plot of V versus d_1 and d_2 can be represented by a contour map of the three dimensional potential energy surface in which V is plotted vertically out of the plane of the paper and the contours are sections of the surface cut by horizontal planes drawn at convenient intervals along the V axis. Figure 3.15 shows a typical contour map for a

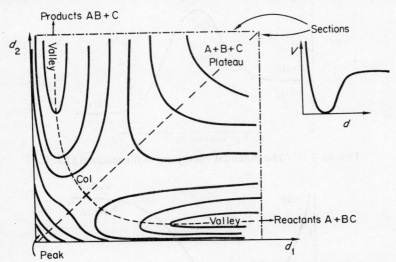

FIGURE 3.15   A contour map of the potential energy surface for the linear reaction [3.36]. Reaction coordinate-----.

reaction such as [3.36]. Vertical sections of this surface normal to the d_2 and d_1 axes at very large values of d_2 and d_1 respectively where the atom and diatomic molecule are separate entities give potential energy curves similar to that of Figure 3.13 for the lowest energy state of a diatomic molecule which can be well represented by the empirical Morse equation,

$$U = D(1 - \exp[-a(d - d_e)])^2 \qquad (3.101)$$

where D, a, d_e are empirical constants.

The potential energy surface has a very high region close to the d_1 and d_2 axes separated by two valleys from a high plateau at large d_1 and d_2 which corresponds to the dissociated atoms $A + B + C$. At the heads of these valleys and separating them there is a col or saddle point. The path of minimum potential energy between the reactant valley ($A + BC$) and product valley ($AB + C$) is shown on Figure 3.15 and passes over the col. This path is called the *reaction coordinate* since it is the path of minimum energy and, therefore, maximum probability (other factors being equal) between reactants and products. To treat the atomic motion in terms of the classical motion of a frictionless mass over this system of hills and valleys it is necessary to skew the coordinate axes and distort their scales, but this does not alter the topological features of the surface

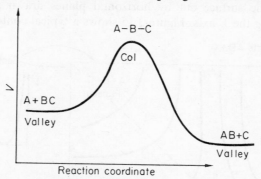

FIGURE 3.16   The potential energy curve for reaction [3.36].

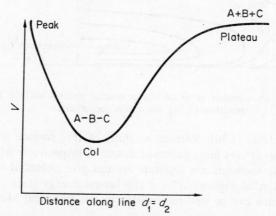

FIGURE 3.17   A plot of potential energy along the line $d_1 = d_2$ for the linear reaction [3.36] in the case that A, B and C are identical.

depicted in Figure 3.15. For many-dimensional potential energy hypersurfaces the general features are also similar to this simple case.

The potential energy V can be plotted as a function of distance along the reaction coordinate as shown in Figure 3.16. This is called the potential energy *curve* or *profile* for the reaction. Along the reaction coordinate the col is at a maximum in potential energy, but in all orthogonal directions it is a minimum. This is illustrated in Figure 3.17 where V is plotted as a function of distance along the line $d_1 = d_2$ for the symmetrical case $A \equiv B \equiv C$.

The surface considered so far is simply that of the lowest energy state (ignoring zero point energy), above this will lie a whole overlapping manifold of hypersurfaces for the different vibrational, rotational and electronic energy levels of the system. The lowest energy the actual molecular system can possess must be at least an amount equal to the zero point vibrational energy above the lowest potential surface represented in Figure 3.15. If the higher electronic and rotational energy levels are not of great importance then a single potential energy surface can be considered for simplicity. This approximation is usually valid but exceptions do occur, e.g. when rotational energy is important (see p. 146) or when the 'crossing' of potential energy surfaces of different electronic levels occurs near the col, see section 3.9.

3.7c The calculation of potential energy surfaces

In practice the calculation of potential energy surfaces by the quantum mechanical method outlined at the beginning of this section is too difficult to be made with any useful degree of accuracy. Even for the simplest chemical reaction [3.37], which has been the subject of literally dozens of such calculations, the values for the height of the col above the reactant valley (the barrier height) that are calculated vary widely (e.g. from 14 to 30 kcal) depending on the approximations used. The experimental value is about 8 kcal and is probably correct to within 1 kcal or so. The reason for the difficult nature of this calculation is that the energy is obtained from the difference between two very large values corresponding to the total ionization of the reactants and the species at the col. At 300°K a change of only 1·4 kcal in the barrier height changes the predicted reaction rate by a factor of 10. This energy difference is only about 0·06 eV. The electronic energies from which the barrier height is calculated are typically of the order of 50 eV. The calculations need to be accurate to within about 0·1% to give the rate constant within an order of magnitude. Such accuracy for the methods of approximation used in quantum mechanical calculations is unobtainable at present. More accurate potential energy surfaces can be obtained by semi-empirical methods in which experimental information obtained for example spectroscopically is used together with quantum mechanical formulae containing adjustable parameters to fit the experimental results. Purely empirical methods have also been used. For example by rotating a Morse function (3.101) for a diatomic molecule through 90° and then distorting the surface so generated, potential energy surfaces in reasonable agreement with experiment have been obtained.

The 'Bond Energy Bond Order' (BEBO) method also relies on an empirical relation, of Pauling,

$$d_n = d_1 - 0 \cdot 6 \log_{10} n \qquad (3.102)$$

together with,

$$V_n = V_1 n^p \qquad (3.103)$$

where d_n is the length (in Å) and V_n is the energy of a bond of order n, and p is an empirical exponent ($p = 1 \cdot 195$ for C–H bonds, $p = 1 \cdot 086$ for H–H bonds). By assuming that along the reaction coordinate the total bond order stays constant, that is the bond order gained by the A–B bond is equal to that lost by the weakened B–C bond, the potential energy curve (but not the complete surface) for the reaction can be calculated and hence the barrier height obtained.

The manner in which the potential energy surface is used to calculate the rate of reaction must now be considered. In section 3.8 the transition state method is described. Here we will briefly outline one collision based treatment of reaction rates using potential energy surfaces.

3.7d Monte Carlo kinematic trajectory calculations

This is essentially a form of theoretical bagatelle, in which frictionless massive particles are shot onto the potential energy surface in a random fashion and the probability of reaction is deduced from the fraction of particles that find their way into the product valley. Two possible particle trajectories are indicated in Figure 3.18.

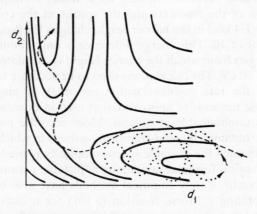

FIGURE 3.18 Examples of trajectories for un-reactive, ········, and reactive, ----, collisions for the linear reaction [3.36].

These trajectories can be calculated by using classical mechanics for simplicity. Starting with randomly chosen initial conditions (by using suitable distribution functions from which to make the choices) and setting the potential energy equal to that for the reactants (large d_1) the equations of motion may be solved to predict where the representative point gets to after a short time interval δt. The new potential energy V for this point is then obtained from the potential energy surface and used to calculate the position reached after the next δt. In this way we can proceed along the trajectory in short segments until either reaction occurs or the reactants fly apart again. By using a computor to repeat this calculation for a large number of trajectories starting from randomly chosen initial conditions the probability of reaction P can then be approximated by

$$P \simeq \frac{\text{no. of trajectories reaching product valley}}{\text{total no. of trajectories calculated}} \qquad (3.104)$$

The variation of P with energy and scattering angles, etc., can also be investigated in this way and hence P_{ij}^{kl} $(E, b, \beta, \theta, \phi, K)$ found. Detailed information about the way in which the energy of the products is distributed is also obtained. This technique has only been applied, as yet, to a few simple reactions such as [3.37] and,

$$CH_3I + K \rightarrow K^+I^- + CH_3^{\cdot} \qquad [3.38]$$

for which detailed experimental evidence for the differential cross-sections as functions of E, θ, ϕ, etc., has been obtained for comparison (see section 2.5c).

This comparison can be used to get information about the potential energy surface of the reaction by trial and error. For reactions such as [3.38] it has been found from crossed molecular beam experiments that most of the exothermicity of the reaction appears in the internal vibrational modes of the products. From this it may be deduced that the potential energy surface is of the 'attractive' type illustrated in Figure 3.19a in which a typical trajectory is plotted. In contrast, infra-red chemiluminescence studies of reaction such as,

$$H + X_2 \rightarrow HX + X \qquad [3.39]$$

(where, X = halogen) have shown that very little of the energy released in the reaction appears as vibration of the new bond, but most is converted into relative translational motion of the products. The type of potential energy surface in this case is called 'repulsive' and is illustrated in Figure 3.19b with a typical trajectory.

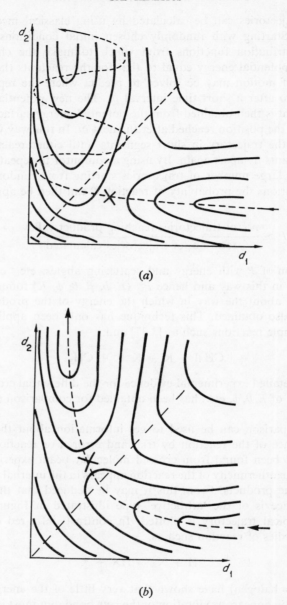

(a)

(b)

FIGURE 3.19 (a) An 'attractive' potential energy surface, (b) a 'repulsive' potential energy surface.

We now turn to a method of using potential energy surfaces that, in contrast, has found almost universal application.

TRANSITION STATE THEORIES

3.8   DERIVATION OF THE RATE EQUATION

This type of approach to the theory of reaction rates largely follows the pioneering work of Eyring. Attention is focussed primarily on the col of the potential energy surface—the transition state—that lies somewhere between reactant and product states. Since the reaction coordinate is the path of maximum probability between reactants and products the total rate of reaction may be approximated by the rate of passage through a region in the vicinity of the transition state. If the potential energy increases rapidly away from the col along directions orthogonal to the reaction coordinate (see Figure 3.17) most reaction will occur by paths passing close to the transition state and the approximation will be a good one. The transition state is treated as a thermodynamic entity (even though it is mechanically unstable) and the reaction rate is evaluated in terms of the properties of this state. These properties are to be obtained preferably from the calculated potential energy surface but more likely by intelligent guesswork. This procedure circumvents the problem of how the reactants enter the transition state. The rate is formulated as the product of the concentration of species in the near vicinity of the transition state (so called activated complexes) and their 'frequency of decomposition', i.e. their average reciprocal lifetime. The concentration of complexes is calculated in terms of the equilibrium concentration that would exist if there were no net reaction which in turn is calculated by normal statistical mechanical methods (see Appendix A).

3.8a   The 'equilibrium hypothesis'

Suppose that it is possible to define a small region at the top of the potential energy barrier of Figure 3.16 such that all systems entering from the left pass through to products without reflexion and similarly all systems entering from the right must pass through to reactants. This might be done by making the region small enough. Species within this region of no return are called 'activated complexes'. Consider now a system in which the reactants are in equilibrium with the products. There will be two types of activated complexes in this system, those moving from reactants to products and those moving in the opposite direction from products to reactants. Denoting the concentrations of these by C_f^* and C_b^* respectively

(f and b standing for forward and backward) and using the prefix e to denote equilibrium values we may write

$$eC_f^* = eC_b^* = \tfrac{1}{2}eC_{f+b}^* \tag{3.105}$$

where suffix $f + b$ refers to both sorts of activated complexes, since it is reasonable to assume that *on average* complexes are equally likely to be moving forward as backward. Since at equilibrium the rate forward equals the rate backward the assumption is equivalent to saying that both sorts of complex have the same average reciprocal lifetime or decomposition frequency. If now the products are instantaneously removed from this equilibrium system,

$$C_b^* = 0 \tag{3.106}$$

but since the forward moving complexes come from reactants only (assuming complexes cannot be reflected back once they have passed through the transition state) then providing the two types of complex do not interact in any way C_f^* should not change. That is, the concentration of forward moving complexes *is the same as it would be in an equilibrium system*

$$C_f^* = eC_f^* = \tfrac{1}{2}eC_{f+b}^* \tag{3.107}$$

This is the 'equilibrium hypothesis' of transition state theory. True equilibrium between reactants and activated complexes is *not* assumed (e.g. see p. 137). Using (3.107) C_f^* may be calculated since eC_{f+b}^* can be obtained using statistical mechanics and the properties of the transition state.

3.8b Calculation of the rate constant

Consider an elementary reaction,

$$A + B + \cdots \rightarrow * \rightarrow \text{products} \tag{3.40}$$

where $*$ is the activated complex situated in a small region at the top of the potential energy barrier, as in Figure 3.20.

Then,

$$\frac{eC_{f+b}^*}{C_A C_B \ldots} = K_c^* \tag{3.108}$$

where K_c^* is the equilibrium constant for the hypothetical equilibrium,

$$A + B + \cdots \rightleftharpoons * \tag{3.41}$$

Using equation (A.62) for the equilibrium constant

$$\frac{eC_{f+b}^*}{C_A C_B \ldots} = \frac{Q^*}{Q_A Q_B \ldots} e^{-E_0^*/RT} \tag{3.109}$$

where Q^* is the partition function per unit volume for the activated complex, (treating this as though it were a stable thermodynamic entity), Q_A, Q_B are those for the reactants, and E_0^* is the energy (per mole) of the lowest level of the activated complex relative to the lowest level of the reactants, see Figure 3.20. The Δ has been dropped from E_0^* as a reminder

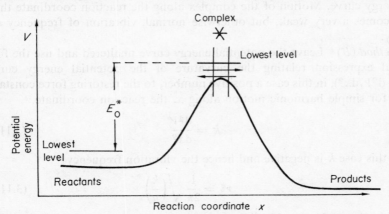

FIGURE 3.20 A potential energy curve showing the 'complex' region and the energy E_0^*.

that this is not strictly a normal thermodynamic quantity since part of the energy of the complex (that relating to motion along the reaction coordinate) is not well defined.

Hence from (3.107) and (3.109),

$$C_f^* = \tfrac{1}{2}eC_{f+b}^* = \tfrac{1}{2}C_AC_B \ldots \frac{Q^*}{Q_AQ_B\ldots} e^{-E_0^*/RT} \qquad (3.110)$$

Now all the partition functions are calculable as outlined in Appendix A except for that part of Q^* which refers to motion along the reaction coordinate. Assuming that this motion is strictly separable from the other degrees of freedom we can write

$$\varepsilon^* = \varepsilon^\ddagger + \varepsilon_{rc}^* \quad \text{per molecule} \qquad (3.111)$$

and hence,

$$Q^* = Q^\ddagger q_{rc}^* \qquad (3.112)$$

where $\varepsilon^\ddagger$ is the energy of the complex in all the normal degrees of freedom and ε_{rc}^* is that in the reaction coordinate, and similarly $Q^\ddagger$ is the partition function for all the $(3N - 7)$ normal degrees of freedom of a non-linear complex ($3N - 6$ for a linear one) and q_{rc}^* is the partition function for motion along the reaction coordinate. $Q^\ddagger$ contains only normal terms since the complex is mechanically stable (i.e. at a potential energy minimum) along all coordinates but the reaction coordinate, see Figures 3.16 and 3.17. q_{rc}^* may be calculated by a number of methods of varying degrees of credibility. We will consider just two.

Method (A)   Imagine a slight depression to exist at the top of the potential energy curve. Motion of the complex along the reaction coordinate then becomes a very weak, but otherwise normal, vibration of frequency ν^*, say.

Method (B)   Leave the potential energy curve unaltered and use the formal expression relating the curvature of the potential energy curve $(-d^2V/dx^2)$, in this case a positive number, to the restoring force constant k, for simple harmonic motion along x, the reaction coordinate.

$$k = \frac{d^2V}{dx^2} \tag{3.113}$$

In this case k is negative and hence the vibration frequency

$$\nu^* = \frac{1}{2\pi}\sqrt{\left(\frac{k}{\mu}\right)} \tag{3.114}$$

is imaginary. Ignore this difficulty and proceed normally.

For a vibrational mode, as shown by (A.41)

$$q_v = (1 - e^{-h\nu/kT})^{-1} \tag{3.115}$$

Where a small q is used to emphasize that this is only part of the whole partition function Q. If the vibration is very weak, i.e. $h\nu \ll kT$, corresponding to a very flat potential energy curve or high temperature, then, from (3.115) by expanding the exponential in powers of $h\nu/kT$,

$$q_v = \frac{kT}{h\nu} \tag{3.116}$$

to a good approximation.

Thus we may take

$$q_{rc}^* = \operatorname*{Lim}_{\nu^* \to 0} (1 - e^{-h\nu^*/kT})^{-1}$$

$$= \frac{kT}{h\nu^*} \tag{3.117}$$

Substitution into (3.110) gives,

$$C_f^* = \tfrac{1}{2}C_A C_B \ldots \frac{(kT/h\nu^*)Q^\ddagger}{Q_A Q_B \ldots} e^{-E_o^*/RT} \tag{3.118}$$

Note that, formally, according to method (B) above, since ν^* is imaginary then so is the concentration of activated complexes C_f^*.

Now the time of transit through the activated complex region is clearly $\frac{1}{2}$ of a vibration period [using either method (A) or (B)]. Thus the average frequency of decomposition of the forward moving complexes is $2\nu^*$ being the reciprocal of the time of transit. The rate of reaction per unit volume r is given by the rate of passage through the activated complex region provided most trajectories leading to products pass close to the col. Thus,†

$$r = \frac{\text{concentration of complexes}}{\text{average time spent in the 'complex' region}} \tag{3.119}$$

$$r = \text{concentration of complexes} \times \text{decomposition frequency} \tag{3.120}$$

Hence, $$r = C_f^* \times 2\nu^* \tag{3.121}$$

$$= C_A C_B \ldots \left(\frac{kT}{h}\right) \frac{Q^\ddagger}{Q_A Q_B \ldots} e^{-E_0^*/RT} \tag{3.122}$$

The rate constant, $k = r/C_A C_B \ldots$

$$\therefore \qquad k = \left(\frac{kT}{h}\right) \frac{Q^\ddagger}{Q_A Q_B \ldots} e^{-E_0^*/RT} \tag{3.123}$$

This result therefore fortunately turns out to be independent of the awkward factor ν^*. Thus in method (A) the depression in the potential energy surface can be decreased to zero in the limit so that,

$$\nu^* \to 0$$

and $$E_0^* \to E_0^\ddagger \qquad \text{from (3.111)}$$

where, $$E_0^* = E_0^\ddagger + \tfrac{1}{2}h\nu^* \tag{3.124}$$

and the rate constant,

$$k = \left(\frac{kT}{h}\right) \frac{Q^\ddagger}{Q_A Q_B \ldots} e^{-E_0^\ddagger/RT} \tag{3.125}$$

In method (B) the product of an imaginary concentration with an imaginary frequency yields a real number for the rate constant since ν^* has cancelled out giving the same result as by method (A) when $E_0^* \to E_0^\ddagger$.

The motion along the reaction coordinate may also be treated as a completely free translation which is equivalent to assuming a perfectly flat

† This may be visualized in terms of particles flowing down a tube or motor cars passing through a bottleneck. If there are on average 100 cars in the bottleneck and each on average takes 10 minutes to pass through it, the rate of traffic flow will be 10 cars per minute through the bottleneck.

potential function (no curvature either positive or negative). This method also leads to the same result (3.125). That (3.125) is independent of the exact shape of the potential curve in the region of the col and whether it has a small positive or small negative or accurately zero curvature makes no difference, gives some reason for having confidence in the result (3.125) in spite of the rather artificial methods used here in its derivation. It is important to remember that $Q^{\ddagger}$ in (3.125) is *not* the partition function of the activated complexes, ∗, but is that of a *hypothetical complex*, ‡, which has lost one degree of freedom, that of the reaction coordinate. In other words in the complex, ‡, motion along the reaction coordinate is frozen or as it were 'clamped', so that it is a completely stable entity, but it is not one that can ever exist in any experimental system even for the most fleeting moment. Whereas, the unstable complex, ∗, does actually exist transitorily in an experimental reaction system, but has ill-defined thermodynamic properties as it is mechanically unstable and is continuously moving towards the products.

Equation (3.125) can be applied to a reaction of *any* molecularity. This is because, owing to the equilibrium hypothesis, the details of the mechanics of activation are immaterial to the final rate expression.

3.8c Symmetry

The expressions derived for the partition function Q in Appendix A were calculated for molecules with no rotational symmetry. When equilibrium constants are calculated from equation (A.62) it is necessary to take account of rotational symmetry elements in the molecules by including the appropriate symmetry numbers (σ) in the rotational partition functions. For example the expression

$$q_r^2 = \frac{8\pi^2 IkT}{h^2} \tag{3.126}$$

is correct for a heteronuclear diatomic molecule (or unsymmetrical linear molecule) but for a homonuclear diatomic (or symmetrical linear molecule) the corresponding expression is

$$q_r^2 = \frac{8\pi^2 IkT}{2h^2} \tag{3.127}$$

This difference occurs because the symmetry laws of quantum mechanics require that such molecules may only occupy odd numbered or even numbered rotational states, but not both, depending on the symmetry of the other parts of the total wave function, (e.g. the nuclear spin wave function whose symmetry depends on whether ortho or para nuclear spin states are considered). At normal temperatures where very many rotational levels are occupied the sums (A.19) over odd and even states are equal and, therefore, equal to one half the sum over both. If partition functions including symmetry numbers are used in (3.125) then this may lead to an incorrect rate expression. The correct procedure when using (3.125) to calculate the rate constant for a reaction involving species with some symmetry is to omit the symmetry numbers entirely and use the partition function expressions for unsymmetrical molecules given in Appendix A, but to multiply the rate constant expression by a statistical factor S thus,

$$k = S \left(\frac{kT}{h}\right) \frac{Q^{\ddagger}}{Q_A Q_B \ldots} e^{-E_0^{\ddagger}/RT} \tag{3.128}$$

S is the number of different activated complexes (or sets of reaction products*) that can be formed if all identical atoms in the reactants are labelled to distinguish them. For example, for the reaction

$$H + H_2 \rightarrow (H \ldots H \ldots H)^{\ddagger} \rightarrow H_2 + H \tag{3.42}$$

$S = 2$ since either end of the H_2 molecule may be attacked; but

$$\sigma_{H_2} = \sigma_{H_3}^{\ddagger} = 2, \quad \text{i.e.} \quad \frac{\sigma_{H_2}}{\sigma_{H_3}^{\ddagger}} = 1.$$

Clearly common sense dictates that (3.128) with $S = 2$ is the correct rate and (3.125) gives a rate too low by a factor of two if partition functions including symmetry numbers are used. The necessity for this modification emphasizes the difference between a true equilibrium and the quasi-equilibrium of transition state theory. In true equilibrium the concentrations of products depend both on their rate of formation and on their rate of reaction back again. In quasi-equilibrium the concentration of activated complexes is determined solely by their rate of formation from the reactants and does not depend on the purely hypothetical rate at which they would return to reactants in an equilibrium system.

3.9   THE TRANSMISSION COEFFICIENT

On page 127 it was assumed that only one electronic state, that of lowest energy, was important in determining the reaction rate. In some cases the potential energy surfaces for two different electronic states 'cross' in the region of the col. In this case not every activated complex (as defined on page 131) that is formed goes on to form products, some are reflected back to reactants, or possibly in more complicated cases to some other set of

* For *acceptable* complexes these should be equal.

products. Two extreme cases may be distinguished as illustrated in Figure 3.21.

In case (a) the curves only cross in a very poor approximation; this occurs with electronic states of the same species. The interaction of the two states in the region of low-order-approximation crossing is large and consequently the 'repulsion' of the two curves is also large. The probability of transition between the lower and upper states (which depends on the overlap of the wave functions) is, therefore, very small and the probability that an activated complex will remain in the lower state and pass over the

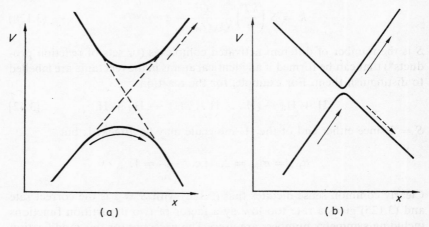

FIGURE 3.21   The 'crossing' of potential energy curves. (a) States of the same species, $\kappa \simeq 1$. (b) States of different species, $\kappa \simeq 0$.

col to reactants is very close to unity, i.e. the transmission coefficient $\kappa \simeq 1$. In the second case, illustrated in (b), the electronic states are of different species and hence interact only in the very near vicinity of the 'crossing point'. The splitting between the true levels at this point is only very small so that the curves cross to quite a high-order-approximation. In this case the probability of transition between the lower and upper states is close to unity and the transmission coefficient, κ, is, therefore, very small. The transition probability depends among other things on the velocity with which the system passes through the crossing region, (high velocities favouring transition between lower and upper levels) so the concept of a fixed transmission coefficient is an approximation. However, the *average* effect of it may be included in the rate equation thus:

$$k = \kappa S \left(\frac{kT}{h}\right) \frac{Q^{\ddagger}}{Q_A Q_B \dots} e^{-E_0^{\ddagger}/RT} \qquad (3.129)$$

where κ is normally unity, but may be small in cases such as (b) above. κ has, on occasion, been used simply as an empirical correction factor to obtain agreement with experiment in a somewhat similar manner to the use of the steric factor P in S.C.T.

A special case of (b) occurs when the two states have different multiplicity (i.e. when reaction is accompanied by reversal of electron spin). The low transmission coefficient to be expected in this case is the basis for the 'spin conservation' rule applied to chemical reactions, according to which spin changes do not occur readily. Direct experimental evidence for the applicability of this rule to normal chemical systems involving atoms of moderate atomic number is hard to find. It is often accepted almost as axiomatic. For example, when interpreting the stereospecificity of addition of methylene to olefines

$$R_2 \underset{R_2}{\overset{R_1}{\diagdown}} C = C \underset{R_4}{\overset{R_3}{\diagup}} + CH_2 \rightarrow R_2 - \overset{R_1}{\underset{\underset{\underset{H}{\diagup} \underset{H}{\diagdown}}{C}}{C}} - C - R_4 \qquad [3.43]$$

it has been used to distinguish between singlet methylene, which it is supposed can 'clip on' to the double bond without multiplicity changes and hence retain the original configuration of the olefine, and triplet methylene, which it is supposed cannot do so thus leading to non-retention of configuration via a biradical intermediate. Experimental evidence for these conclusions is dubious and the subject of conflicting interpretations. The production of methylene by pyrolysis of azomethane

$$CH_2N_2 \rightarrow CH_2 + N_2 \qquad [3.44]$$

is assumed to lead to singlet methylene owing to the spin conservation rule. This multiplicity may then be used to interpret the relative stereospecificity of the reactions of CH_2 derived from a variety of sources to provide evidence for spin conservation. This circular argument is further invalidated by the fact that the corresponding reaction to [3.44] of the isoelectronic molecule nitrous oxide,

$$\underset{(^1\Sigma)}{N_2O} \rightarrow \underset{(^1\Sigma)}{N_2} + \underset{(^3P)}{O} \qquad [3.45]$$

occurs contrary to the spin conservation rule with a frequency factor of 8×10^{11} sec^{-1} which is equal to that for [3.44] within experimental error. The transmission coefficient for [3.45] has been calculated to be about 10^{-1} to 10^{-2} by estimating the splitting of the potential energy curves due to spin–orbit interaction as about 500 cal. Such a factor is relatively small compared to the effects of other variables on the rates of chemical reactions. The rule should therefore be applied with caution.

3.10 BIMOLECULAR REACTIONS

Although the bimolecular reaction of two atoms is impossible for reasons outlined earlier (p. 96) it is interesting to compare S.C.T. with T.S.T. by applying the latter theory to the hard sphere bimolecular collision model of the former,

$$A + B \rightarrow (AB)^{\ddagger} \rightarrow \text{products} \qquad [3.46]$$

The atoms A and B have only translational degrees of freedom, while the complex $(A - B)^{\ddagger}$ has both translational and rotational motion (but no vibration since the A–B distance is fixed in the $\ddagger$ complex). Hence,

$$Q_A = \left(\frac{2\pi m_A kT}{h^2}\right)^{\frac{3}{2}} \quad \text{and} \quad Q_B = \left(\frac{2\pi m_B kT}{h^2}\right)^{\frac{3}{2}}$$

and

$$Q^{\ddagger} = \left(\frac{2\pi(m_A + m_B)kT}{h^2}\right)^{\frac{3}{2}} \cdot \frac{8\pi^2 IkT}{h^2}$$

where the moment of inertia $I = \mu d_{AB}^2$ and d_{AB} is the distance apart of the centres of the atoms in the complex.

The bimolecular rate constant is from (3.125)

$$
{}_2k = \frac{\left(\dfrac{kT}{h}\right)\left(\dfrac{2\pi(m_A + m_B)kT}{h^2}\right)^{\frac{3}{2}} \dfrac{8\pi^2 \left(\dfrac{m_A m_B}{m_A + m_B}\right) d_{AB}^2 kT}{h^2} e^{-E_0^{\ddagger}/RT}}{\left(\dfrac{2\pi m_A kT}{h^2}\right)^{\frac{3}{2}} \left(\dfrac{2\pi m_B kT}{h^2}\right)^{\frac{3}{2}}}
$$

$$= \pi d_{AB}^2 \left(\frac{8kT}{\pi\mu}\right)^{\frac{1}{2}} e^{-E_0^{\ddagger}/RT} \qquad (3.130)$$

which is the same result as S.C.T. (3.22) provided that

$$d_{AB} = \sigma_{AB} \qquad (3.131)$$
$$E_0^{\ddagger} = E_c \qquad (3.132)$$

These are both reasonable requirements in view of the definitions of these quantities. (3.131) is clearly true for hard spheres, while in (3.132) the critical kinetic energy of relative motion along line of centres E_c is equal to the height of the potential energy barrier to be surmounted $E_0^{\ddagger}$ for classical motion.

This is the only case in which the two theories give the same result. In general for bimolecular reactions of molecules, Q_A, Q_B and $Q^{\ddagger}$ contain contributions from all the rotational and vibrational motions of these species and so calculation for large molecules becomes quite complicated. If only very approximate rate constants are required, within a few orders of magnitude, say, the approximation can be made that the partition functions per degree of translational or rotational or vibrational freedom are the same for all species, i.e.

$$q_t = \left(\frac{2\pi mkT}{h^2}\right)^{\frac{1}{2}} \simeq \text{constant}$$

$$q_r = \left(\frac{8\pi^2 IkT}{h^2}\right)^{\frac{1}{2}} \simeq \text{constant}$$

$$q_v = (1 - e^{-h\nu/kT})^{-1} \simeq \text{constant}$$

The first two approximations are accurate if the masses and moments of inertia are all approximately equal, while the last one is reasonable provided either that the vibration frequencies are the same or, as is more likely, that the temperature is sufficiently low that $kT \ll h\nu$ and hence q_v is close to unity. Using this crude approximation for a non-linear molecule

$$Q = q_t^3 q_r^3 q_v^{(3N-6)} \tag{3.133}$$

and for a linear one

$$Q = q_t^3 q_r^2 q_v^{(3N-5)} \tag{3.134}$$

For the complex, $\ddagger$, $Q^{\ddagger}$ has one fewer vibrational degree of freedom than the corresponding molecule. At normal temperatures 'typical' values for the partition functions per degree of freedom are:

$$\left.\begin{array}{l} q_t \simeq 10^8 - 10^9 \text{ cm}^{-1} \\ q_r \simeq 10^1 - 10^2 \\ q_v \simeq 1 - 10 \end{array}\right\} \tag{3.135}$$

In the case of two atoms considered above, using these approximate results

$$_2k = \left(\frac{kT}{h}\right) \frac{q_t^3 q_r^2}{q_t^3 q_t^3} e^{-E_0{}^{\ddagger}/RT} \tag{3.136}$$

which by (3.130)

$$= Z_{AB}\, e^{-E_0{}^{\ddagger}/RT} \tag{3.137}$$

Expressing differences between this S.C.T. result and other cases in terms of a 'steric factor' P

$$P = {}_2k/Z_{AB}\, e^{-E_0^\ddagger/RT} \tag{3.138}$$

we may calculate P in terms of the q's and hence obtain approximate order of magnitude values for P for the various types of bimolecular reactions as in Table 3.2.

TABLE 3.2   Order of magnitude steric factors calculated from T.S.T. for bimolecular reactions of atoms, A, linear molecules, L, and non-linear molecules, N. n is the exponent of temperature in the pre-exponential factor.

Reactants	Complex	Steric factor P	Order of magnitude of P	n
A + A	L	1	1	$+\frac{1}{2}$
A + L	L	$(q_v/q_r)^2$	10^{-2}	$-\frac{1}{2}$
A + L	N	(q_v/q_r)	10^{-1}	0
A + N	N	$(q_v/q_r)^2$	10^{-2}	$-\frac{1}{2}$
L + L	L	$(q_v/q_r)^4$	10^{-4}	$-\frac{3}{2}$
L + L	N	$(q_v/q_r)^3$	10^{-3}	-1
L + N	N	$(q_v/q_r)^4$	10^{-4}	$-\frac{3}{2}$
N + N	N	$(q_v/q_r)^5$	10^{-5}	-2

For example, taking the last line in the table, for the reaction between two non-linear molecules containing N_A and N_B atoms respectively via a non-linear complex, the rate expression (neglecting statistical and transmission factors) is;

$$\begin{aligned}
{}_2k &= \left(\frac{kT}{h}\right)\frac{q_t^3 q_r^3 q_v^{3(N_A+N_B)-7}}{q_t^3 q_r^3 q_v^{3N_A-6}\, q_t^3 q_r^3 q_v^{3N_B-6}}\, e^{-E_0^\ddagger/RT} \\[2mm]
&= \frac{kT}{h}\left(\frac{q_t^3 q_r^2}{q_t^3 q_t^3}\right)\left(\frac{q_r q_v^{12}}{q_r^6 q_v^7}\right) e^{-E_0^\ddagger/RT} \\[2mm]
&= Z_{AB}(q_v/q_r)^5 e^{-E_0^\ddagger/RT} \qquad \text{by (3.136) and (3.137)}
\end{aligned}$$

Hence $P = (q_v/q_r)^5$ by (3.138).

The partial partition functions q are functions of temperature. Thus

$$\left.\begin{aligned}
q_t &\propto T^{\frac{1}{2}} \\
q_r &\propto T^{\frac{1}{2}} \\
\text{while at low } T, \; q_v &\propto T^0 \text{ approximately.}
\end{aligned}\right\} \tag{3.139}$$

Hence in general, we have approximately

$$_2k \propto T^n \, e^{-E_0^\ddagger/RT} \qquad (3.140)$$

Where $\pm n$ is integral or half-integral. Values of n are given in Table 3.2. From (3.140) the experimental activation energy is

$$_2E_a = -\frac{\mathrm{d}\ln_2k}{\mathrm{d}(1/RT)}$$

$$= E_0^\ddagger + nRT \qquad (3.141)$$

The temperature dependence of $_2E_a$ predicted is therefore rather small for bimolecular reactions since $-2 \leqslant n \leqslant +\tfrac{1}{2}$. This agrees with experiment.

The order of magnitude results in Table 3.2 afford a ready interpretation of the observed decrease in steric factors for elementary bimolecular reactions with increasing complexity of the reactants discussed earlier on p. 92. However, such approximate calculations are of very limited use, and in general more accurate values of the partition functions must be obtained from the molecular constants (masses, moments of inertia and vibration frequencies). In the case of the activated complex, the evaluation of $Q^\ddagger$ presents some difficulty since no direct experimental data can exist for this hypothetical non-existent species. The choice of suitable values for the geometrical parameters and vibration frequencies of the complex is often a matter for inspired guesswork and analogy with stable molecules. Often the procedure is reversed and an experimental value for the rate constant is used to deduce some information concerning the nature of the activated complex. Accumulation of such empirical data then facilitates choice of parameters in other systems where experimental evidence is lacking. More detailed examples of the application of T.S.T. to bimolecular reactions will not be considered owing to this uncertainty of choice of complex parameters.

3.11 TERMOLECULAR REACTIONS

Since the mechanism of formation of the activated complex is immaterial in T.S.T. owing to the 'equilibrium' hypothesis, the question of whether any third-order reactions are genuine termolecular processes does not arise. T.S.T. cannot be used to distinguish between the various possible detailed mechanisms for 'termolecular' processes (see section 3.3) nor to interpret any experimentally observed effects that are attributable to differences in these mechanisms. We may, therefore, write

$$A + B + C \rightarrow (ABC)^\ddagger \rightarrow \text{products} \qquad [3.47]$$

without implying anything about the mechanism or molecularity of the reaction. For example, consider the nitric oxide–oxygen reaction,

$$2NO + O_2 \rightarrow \begin{pmatrix} O\cdots\cdots O \\ \vdots \qquad \vdots \\ N \qquad N \\ \| \qquad \| \\ O \qquad O \end{pmatrix}^{\ddagger} \rightarrow 2NO_2 \qquad [3.48]$$

Using order of magnitude partition functions and assuming that the complex has a free internal rotation about the O–O bond rather than a vibration, the rate constant is, from (3.125)

$$_3k = \left(\frac{kT}{h}\right) \frac{Q^{\ddagger}}{Q_{NO}^2 Q_{O_2}} e^{-E_0^{\ddagger}/RT}$$

and,

$$Q^{\ddagger} = q_t^3 q_r^3 q r q_v^{(3\times6-7)-1}$$

$$\therefore \quad _3k = \left(\frac{kT}{h}\right) \frac{q_t^3 q_r^4 q_v^{10}}{q_t^9 q_r^6 q_v^3} e^{-E_0^{\ddagger}/RT}$$

$$= \left(\frac{kT}{h}\right) \frac{q_v^7}{q_t^6 q_r^2} e^{-E_0^{\ddagger}/RT}$$

Taking the values given in (3.135)

$$_3k \simeq 10^{13} \frac{1}{10^{50}10^3} e^{-E_0^{\ddagger}/RT}$$

$$\simeq 10^{-40} e^{-E_0^{\ddagger}/RT} \quad (\text{cc}^2 \text{ molecule}^{-2} \text{ sec}^{-1})$$

$$\simeq 3{\cdot}6 \times 10^7 e^{-E_0^{\ddagger}/RT} \quad (\text{cc}^2 \text{ mole}^{-2} \text{ sec}^{-1})$$

(to within about ± 5 orders of magnitude), which is in reasonable agreement with the experimental value (page 97 and Table 4.3, $P = 10^{-7}$) if $E_0^{\ddagger} = 0$. The temperature coefficient of the rate constant is given from (3.139) as,

$$_3k \propto T^1 T^{-3} T^{-1} e^{-E_0^{\ddagger}/RT} = T^{-3} e^{-E_0^{\ddagger}/RT}$$

Hence the experimental activation energy is

$$E_a = E_0^{\ddagger} - 3RT$$

This, again, fits well with experiment if $E_0^{\ddagger} = 0$. The experimental activation energy is negative and a plot of $_3kT^3$ versus T gives a reasonably good horizontal line at moderate temperatures. By taking reasonable values for the moments of inertia and vibration frequencies of the complex (by analogy with N_2O_4) the above order of magnitude calculation can be improved to give excellent agreement between theory and experiment for $_3k$. The other termolecular chemical reactions may be treated with similar success to this example. Third-order atom and radical recombination reactions cannot be successfully dealt with using T.S.T. for reasons outlined in the next section for the reverse reactions.

3.12   UNIMOLECULAR   REACTIONS

The method used on p. 131 to derive the 'equilibrium' hypothesis of T.S.T. by considering an equilibrium reaction system in which the products were instantaneously removed shows clearly that the hypothesis is only applicable to systems in which the reactant molecules themselves are in thermal equilibrium, that is they are distributed over the available energy levels according to a Boltzmann distribution. This was further implied when partition functions were used to calculate the concentration of complexes. The T.S.T. result (3.125) cannot be used in situations where the energy distribution is non-Boltzmann, such as unimolecular reactions in the fall-off and low pressure regions and their reverses such as atom and radical recombinations which require third bodies. To deal with such reactions 'statistical' assumptions such as that given on p. 111 for the RRKM theories can be used to replace the 'equilibrium' assumption. These effectively assume equilibrium over the internal energy states for isolated molecules of fixed total energy without the distribution of molecular energies themselves being Boltzmann. Replacement of partition functions by energy state densities (p. 116) then enables evaluation of the rate of reaction under conditions of known energy distribution.

The T.S.T. result (3.125) may only be applied to unimolecular reactions at infinitely high pressures, i.e. to the calculation of $_1k^\infty$ in the notation of section 3.6.

Considering the unimolecular reaction

$$A \rightarrow (A)^{\ddagger} \rightarrow \text{products} \qquad [3.49]$$

(3.125) gives,

$$_1k^\infty = \left(\frac{kT}{h}\right)\frac{Q^{\ddagger}}{Q_A}e^{-E_0^{\ddagger}/RT} \qquad (3.142)$$

Now the mass of the complex and reactant are equal so that the translational partition functions cancel out leaving,

$$_1k^\infty = \left(\frac{kT}{h}\right)\frac{\Pi q_r^{\ddagger}\Pi q_v^{\ddagger}}{\Pi q_r^A\Pi q_v^A}e^{-E_0^{\ddagger}/RT} \qquad (3.143)$$

Two extreme cases can be distinguished.

(a) *Rigid complexes.* In this case $(A)^{\ddagger}$ is not very different from A. In particular if the bond lengths are similar

$$I_{\ddagger} = I_{A}$$

for all the axes of rotation and if either the vibration frequencies are the same or else $h\nu \gg kT$ so that $(1 - e^{-h\nu/kT})^{-1} = 1$ for all the reactant and complex vibrations, then (3.143) becomes

$$_1k^{\infty} = \left(\frac{kT}{h}\right) e^{-E_0{}^{\ddagger}/RT} \qquad (3.144)$$

For normal temperatures (e.g. $T \approx 400°K$) the pre-exponential factor

$$\frac{kT}{h} = \frac{1\cdot4 \times 10^{-16} \times 400}{6\cdot6 \times 10^{-27}}$$

$$\approx 10^{13} \text{ sec}^{-1}$$

Many unimolecular reactions do have high pressure pre-exponential factors in this region (see Table 4.1) as has already been pointed out in connexion with the Slater and RRKM theories which both predict a similar result, (pp. 113, 118).

(b) *Loose complexes.* For example, in the dissociation of a molecule such as ethane

$$C_2H_6 \rightarrow 2CH_3^{\cdot} \qquad [3.50]$$

the maximum in the effective potential energy curve occurs at large values of the C–C distance (see Figure 3.13). This rotational energy barrier may be at internuclear distances of up to 10Å or so. With this very considerable bond extension not only is

$$I_{\ddagger} \gg I^A$$

so that,

$$q_r^{\ddagger} \gg q_r^A$$

but also some of the vibrations become very loose rocking motions or almost free rotations owing to the very much reduced restoring forces in the complex. Thus for these modes

$$\nu^{\ddagger} \ll \nu^A$$

so that

$$q_v^{\ddagger} \gg q_v^A$$

The net result may be that $Q^{\ddagger}/Q_A$ is several orders of magnitude greater than unity. For example for [3.50] the experimental pre-exponential factor is about 10^{17} sec^{-1} (Table 4.1) and several detailed treatments along the lines outlined above have predicted values in satisfactory agreement with this very high figure. In this connexion it may be remembered that the reverse reaction gives quite a good fit to S.C.T. (p. 92) so that the methyl radicals behave as spheres there being no orientation requirements. This is consistent with the notion of free rotation and very loose rocking motions in the complex for the reverse dissociation.

3.13 PSEUDO-THERMODYNAMIC FORMULATION OF THE T.S.T. RATE EQUATION

In the above paragraphs we have outlined how T.S.T. may be applied to reactions of ideal gases by making suitable estimates about the activated complex. However the real power of T.S.T. really only becomes apparent when considering the far more difficult problems of reactions in non-ideal systems. For these systems the calculation of partition functions becomes difficult if not impossible owing to the appreciable interactions between the molecules, and the statistical mechanics of interacting assemblies loses much of the simplicity associated with ideal gases. However, thermodynamic descriptions of non-ideal systems are relatively straightforward provided experimental data is used to determine the thermodynamic functions in place of statistical calculation. For these reasons it is frequently convenient to cast the rate expression of T.S.T. (3.125) into a thermodynamic form. Equation (3.125)

$$k = \left(\frac{kT}{h}\right) \frac{Q^{\ddagger}}{Q_A Q_B \dots} e^{-E_0^{\ddagger}/RT}$$

contains a 'universal frequency' $\left(\dfrac{kT}{h}\right)$, and the term

$$\frac{Q^{\ddagger}}{Q_A Q_B \dots} e^{-E_0^{\ddagger}/RT}$$

which *formally* is identical to the statistical mechanical expression (A.62) for the equilibrium constant, $K_c^{\ddagger}$, (in terms of concentrations) for the ideal gas reaction

$$A + B + \cdots \rightleftharpoons (AB \dots)^{\ddagger} \qquad [3.51]$$

The Δ may be replaced in the exponential energy term because the complex, $\ddagger$, (unlike $*$) has a well defined energy since motion along the reaction coordinate is excluded. Thus (3.125) becomes

$$k = \left(\frac{kT}{h}\right) K_c^{\ddagger} \qquad (3.145)$$

The thermodynamic relation for the equilibrium constant $K_c^{\ddagger}$ is,

$$(\Delta G^{\ddagger})_c^{\circ} = -RT \ln K_c^{\ddagger} \qquad (3.146)$$

where $(\Delta G^{\ddagger})_c^{\circ}$ is the difference in the standard state free energies of complex, $\ddagger$, and reactants, standard states being defined in terms of concentrations (not pressures). Substituting (3.146) into (3.145),

$$k = \left(\frac{kT}{h}\right) e^{-(\Delta G^{\ddagger})_c^{\circ}/RT} \qquad (3.147)$$

and by using the constant temperature relation

$$\Delta G = \Delta H - T\Delta S \qquad (3.148)$$

this becomes,

$$k = \left(\frac{kT}{h}\right) e^{(\Delta S^{\ddagger})_c^{\circ}/R} e^{-(\Delta H^{\ddagger})_c^{\circ}/RT} \qquad (3.149)$$

$(\Delta G^{\ddagger})_c^{\circ}$, $(\Delta H^{\ddagger})_c^{\circ}$, and $(\Delta S^{\ddagger})_c^{\circ}$ are called the 'standard free energy, enthalpy and entropy of activation' respectively.

Comparing (3.149) with the experimental Arrhenius equation

$$k = A\, e^{-E_a/RT} \qquad (3.150)$$

and the S.C.T. equation

$$k = PZ\, e^{-E_c/RT} \qquad (3.151)$$

it can be seen that broadly speaking $(\Delta H^{\ddagger})_c^{\circ}$ correlates with E_a and E_c respectively, and

$$\left(\frac{kT}{h}\right) e^{(\Delta S^{\ddagger})_c^{\circ}/R}$$

correlates with A or PZ. If the units used are cc, moles and sec, then unit steric factor corresponds approximately to $(\Delta S^{\ddagger})_c^{\circ} = 0$. Steric factors greater than unity correspond to positive entropies of activation and P factors less than unity correspond to negative entropies of activation.

The exact relations may be derived as follows. The experimental activation energy is defined by,

$$E_a = -\frac{d\ln k}{d(1/RT)}$$

which by (3.145) is,

$$= -\frac{d\ln[(kT/h)K_c^{\ddagger}]}{d(1/RT)}$$

now since $K_c^{\ddagger}$ is a *concentration* equilibrium constant the van't Hoff isochore is from (1.89)

$$-\frac{\mathrm{d}\ln K_c^{\ddagger}}{\mathrm{d}(1/RT)} = (\Delta E^{\ddagger})_c^{\circ} \tag{3.152}$$

and hence

$$E_a = (\Delta E^{\ddagger})_c^{\circ} + RT \tag{3.153}$$

$(\Delta E^{\ddagger})_c^{\circ}$ is the 'standard energy of activation' and care must be exercised to avoid confusion with E_a the 'activation energy'. Now by definition,

$$H = E + PV \tag{3.154}$$

and therefore

$$\Delta H = \Delta E + \Delta(PV) \tag{3.155}$$

For ideal gas reactions at constant temperature (see pp. 28, 29)

$$\Delta(PV) = \Delta n RT \tag{3.156}$$

Hence

$$\Delta H = \Delta E + RT \cdot \Delta n \tag{3.157}$$

and so

$$\Delta H^{\ddagger} = \Delta E^{\ddagger} + \Delta n^{\ddagger} RT \tag{3.158}$$

(The standard state conditions are not needed on $\Delta n^{\ddagger}$ since the change in number of moles in the stoichiometric equation is constant, irrespective of the states of the reactants and products.) We have replaced Δv of Chapter 1 by Δn here to avoid confusion with the frequency v. From (3.158) and (3.153)

$$\Delta H^{\ddagger} = E_a + (\Delta n^{\ddagger} - 1)RT \tag{3.159}$$

Substituting into (3.149) gives

$$k = \left(\frac{kT}{h}\right) \mathrm{e}^{+(\Delta S^{\ddagger})_c^{\circ}/R} \, \mathrm{e}^{-(\Delta n^{\ddagger}-1)} \, \mathrm{e}^{-E_a/RT} \tag{3.160}$$

Comparison with (3.150) gives

$$A = \left(\frac{kT}{h}\right) \mathrm{e}^{(\Delta S^{\ddagger})_c^{\circ}/R} \, \mathrm{e}^{-(\Delta n^{\ddagger}-1)} \tag{3.161}$$

$$\left.\begin{array}{ll} \text{For a unimolecular reaction} & \Delta n^{\ddagger} = 0 \\ \text{For a bimolecular reaction} & \Delta n^{\ddagger} = -1 \\ \text{For a termolecular reaction} & \Delta n^{\ddagger} = -2 \end{array}\right\} \tag{3.162}$$

These together with (3.159) and (3.161) give the exact relations between the Arrhenius parameters A and E_a and the thermodynamic ones $\Delta H^{\ddagger}$ and $(\Delta S^{\ddagger})_c^{\circ}$ for reactions of any molecularity of ideal gases.

In general for systems other than ideal gases (3.149) and (3.155) give

$$k = \left(\frac{kT}{h}\right) e^{(\Delta S^{\ddagger})_c^{\circ}/R} \; e^{-(\Delta E^{\ddagger})_c^{\circ}/RT} \; e^{-P(\Delta V^{\ddagger})_c^{\circ}/RT} \tag{3.163}$$

for reactions at constant pressure P. For reactions in solution $(\Delta V^{\ddagger})_c^{\circ}$ the 'standard volume of activation' is therefore given by,

$$(\Delta V^{\ddagger})_c^{\circ} = -RT \left(\frac{\partial \ln k}{\partial P}\right)_T \tag{3.164}$$

In non-ideal systems it is often convenient to compare the reaction rate with that which would be observed in an ideal system. This may be done in an empirical manner by making use of activity coefficients. Thus, letting the activity a of a species in a non-ideal system be given by

$$a = \gamma C \tag{3.165}$$

where C is its concentration and γ the activity coefficient, the rate of reaction on T.S.T. is still given by the relation (3.121)

$$r = 2v^* C_f^*$$

since this is independent of the ideality or non-ideality of the system, being a simple consequence of the conservation of mass. Also, as before

$$C_f^* = \tfrac{1}{2} e C_{f+b}^* \tag{3.107}$$

since the equilibrium hypothesis holds for non-ideal and ideal systems equally. However, equation (3.108) becomes

$$\frac{e a_{f+b}^*}{a_A a_B \ldots} = K^* \tag{3.166}$$

Substituting (3.165) into (3.166),

$$K^* = \frac{e C_{f+b}^*}{C_A C_B \ldots} \frac{e \gamma_{f+b}^*}{\gamma_A \gamma_B \ldots} \tag{3.167}$$

Hence from (3.121) and (3.107),

$$r = v^* C_A C_B \ldots K^* \frac{\gamma_A \gamma_B \cdots}{e \gamma_{f+b}^*} \tag{3.168}$$

Defining $K^{\ddagger}$ as before by

$$K^{\ddagger} \frac{kT}{h\nu^*} = K^* \tag{3.169}$$

$$\nu^* K^* = \left(\frac{kT}{h}\right) K^{\ddagger}$$

Substituting into (3.168)

$$k = r/C_A C_B \ldots = \left(\frac{kT}{h}\right) K^{\ddagger} \frac{\gamma_A \gamma_B \cdots}{e\gamma_{f+b}^*} \tag{3.170}$$

If it is now assumed that the activity coefficient for the unstable complex $e\gamma_{f+b}^*$, is equal to that for the stable, hypothetical, complex, $\gamma^{\ddagger}$, (3.170) becomes

$$k = \left(\frac{kT}{h}\right) K^{\ddagger} \frac{\gamma_A \gamma_B \cdots}{\gamma^{\ddagger}} \tag{3.171}$$

or from (3.145)

$$k = k_{\text{ideal}} \frac{\gamma_A \gamma_B \cdots}{\gamma^{\ddagger}} \tag{3.172}$$

The use of T.S.T. in non-ideal systems will not be pursued further since it would lead us into the complex field of reactions in solutions, without adding to the general principles outlined above.

Free energy surfaces

Equation (3.147) relates the rate constant to the free energy of activation of the reaction and universal constants only. This suggests that an expression for the rate constant might be derived directly from considerations of the free energy of the system without introducing 'mechanical' concepts such as the potential energy surfaces that are used in T.S.T. to define the transition state and determine its concentration and lifetime.

The position defined as the transition state is the 'point of no return' where motion of the molecules becomes free and spontaneous towards the products and the probability of reflexion back into reactant states becomes zero. For a single complex spontaneity of motion is governed by its potential energy. For large statistical assemblies of complexes the entropy, i.e. probability, must also be taken into account. In this case it is the free energy that determines the direction of spontaneous motion *of the whole system*. It has been suggested that it may be that 'free energy surfaces' are needed to define the transition state (sometimes potential energy diagrams are mis-labelled as free energy diagrams). A free energy

6

surface has not yet been satisfactorily *defined*, let alone constructed, without making use of knowledge of the very rates of processes that T.S.T. sets out to calculate.

SUGGESTIONS FOR FURTHER READING

Textbooks

S. W. Benson, *Foundations of Chemical Kinetics*, McGraw-Hill, New York, 1960.
H. S. Johnston, *Gas-phase Reaction Rate Theory*, Ronald, New York, 1966.
V. N. Kondrat'ev, *Chemical Kinetics of Gas Reactions*, Pergamon, Oxford, 1964.
K. J. Laidler, *Chemical Kinetics*, (2nd Ed.), McGraw-Hill, New York, 1965.
N. B. Slater, *Theory of Unimolecular Reactions*, Methuen, London, 1959.

Reviews

Advan. Photochem., **3**, 1, (1964); *Chem. Soc. Spec. Publ.*, **16**, (1962); *Progr. Reaction Kinetics*, **3**, 1, (1965); *Quart. Rev.*, **14**, 133, (1960); *Quart. Rev.*, **18**, 122, (1964).

CHAPTER 3 PROBLEMS

1. Calculate the rate of bimolecular collisions in a pure gas M.W. = 50 at 1 atmosphere pressure and 500°C, assuming that the molecules are hard spheres of 4Å diameter. If chemical reaction were to occur on every such encounter what would be the half-life of the reaction?

2. The viscosity of hydrogen at 0°C is $8 \cdot 6 \times 10^{-5}$ poise (c.g.s. units). Calculate the collision number Z_{AA} for bimolecular collisions in pure hydrogen at 0°C.

3. What fraction of bimolecular collisions at 500°C have a relative kinetic energy along the line of centres exceeding 50 kcal mole^{-1}?

4. What fraction of bimolecular collisions between non-linear molecules each containing 4 atoms at 500°C have a total energy exceeding 50 kcal mole^{-1}?

5. Use S.C.T. to calculate the pre-exponential factor A for the hypothetical bimolecular reaction $2HI \rightarrow H_2 + I_2$ at 600°K, given that $\sigma_{HI} = 3 \cdot 5$Å.

6. The rate constant of the reaction $C_2H_5^{\cdot} + H_2 \rightarrow C_2H_6 + H$ is $k = 10^{11.8}$ exp$(-11,300/RT)$ cc mole^{-1} sec^{-1}. Calculate the steric factor at 800°K given that $\sigma_{H_2} = 2 \cdot 4$Å and $\sigma_{C_2H_5^{\cdot}} = 3 \cdot 0$Å.

7. The critical relative kinetic energy along line of centres for the reaction $H + H_2 \rightarrow H_2 + H$ is 8 kcal mole^{-1}. Use S.C.T. to calculate the activation energy E_a and pre-exponential factor A at 1000°K given that $\sigma_{H_2} = 2 \cdot 4$Å, $\sigma_H = 1 \cdot 4$Å and $P = 1$. What are the corresponding values at 3000°K?

8. What is the rate of termolecular collisions in a pure gas M.W. = 50 at 1 atmosphere pressure and 500°C if $\sigma = 4\text{Å}$? If chemical reaction were to occur on every such collision what would be the half-life of the reaction?

9. The energy of combination of $I + I_2 \rightarrow I_3$ is $\Delta E° = -4.7$ kcal mole^{-1}. If the critical energy E_c for the reaction $I + I_3 \rightarrow 2I_2$ is zero what is the overall activation energy for the recombination of iodine atoms in the presence of molecular iodine at 25°C?

10. Plot out, as a function of energy, the probability of at least 50 kcal mole^{-1} accumulating in one particular oscillator of a set of 10 loosely coupled simple harmonic oscillators.

11. Suggest a mechanism to explain the results given in problem 8 at the end of Chapter 1 for the decomposition of nitryl chloride. Given that the collision diameter of nitryl chloride is 6.7Å calculate the number of squared terms that contribute to activation of this molecule.

12. The high pressure first-order rate constant for the decomposition of perchloric acid, $HOClO_3 \rightarrow OH^. + ClO_3^.$, is $_1k^\infty = 10^{14} \exp(-48,000/RT)$ sec^{-1}. Take the collision frequency Z to be 10^{14} cc mole^{-1} sec^{-1} and assume that one half of the normal modes of $HOClO_3$ contribute to activation and that the steric factor for activation is unity and hence estimate the pressure (in torr) at which the first-order rate constant falls to one half of $_1k^\infty$ at 1000°K.

13. The decomposition of ethyl radicals, $C_2H_5^. \rightarrow C_2H_4 + H$ has a high pressure first-order rate constant $_1k^\infty = 10^{13} \exp(-40,000/RT)$ sec^{-1}. If one half of the normal modes of $C_2H_5^.$ are active calculate the activation energy that would be observed for the low pressure second-order rate constant at 600°C. If the addition of hydrogen atoms to ethylene has an activation energy at high pressures of 1.9 kcal mole^{-1} what would its activation energy be at low pressures at 600°C?

14. The first-order rate constant for the decomposition of the methoxymethyl radical, $CH_3OCH_2^. \rightarrow CH_3^. + CH_2O$, at 270°C varies with the concentration of inert gas M as follows

$_1k$ (sec^{-1})	7.83	9.40	11.8	13.4
10^6[M] (mole cc^{-1})	0.166	0.286	0.500	0.770

Assume that over this limited range of conditions the simple Lindemann theory is adequate and hence estimate the rate constant for the energy transfer reaction $M + CH_3OCH_2^. \rightarrow M + CH_3OCH_2^{.*}$.

15. Use the BEBO method to obtain the height of the potential energy barrier for the reaction $H + H_2 \rightarrow H_2 + H$, given that the exponent p in the empirical equation $V_n = V_1 n^p$ is 1.086 and that V_1, the potential energy of the H–H bond, (not including zero-point energy) is 109.4 kcal mole^{-1}.

16. Give the statistical factor S for each of the following reactions,

(a) $H + H_2$ $\rightarrow$ $(H \ldots H \ldots H)^{\ddagger}$ $\rightarrow$ $H_2 + H$

(b) $H + HD$ $\rightarrow$ $(H \ldots H \ldots D)^{\ddagger}$ $\rightarrow$ $H_2 + D$

(c) $H + DH$ $\rightarrow$ $(H \ldots D \ldots H)^{\ddagger}$ $\rightarrow$ $HD + H$

(d) $CH_3^{'} + H_2$ $\rightarrow$ $(CH_3 \ldots H \ldots H)^{\ddagger}$ $\rightarrow$ $CH_4 + H$

(e)

$$\begin{array}{c} CH_2 \\ \diagup \quad \diagdown \\ CH_2 \text{---} CH_2 \end{array} \rightarrow \left(\begin{array}{c} H \\ \vdots \\ CH \diagup \begin{array}{c} CH_2 \\ | \\ CH_2 \end{array} \end{array} \right)^{\ddagger} \rightarrow CH_3CH = CH_2$$

(f)

$$\begin{array}{c} CH_2 \\ \diagup \quad \diagdown \\ CH \qquad CH_2 \\ \| \qquad \diagup \\ CH \text{---} CH_2 \end{array} \rightarrow \left(\begin{array}{c} CH_2 \\ \diagup \quad \diagdown \\ CH \qquad CH \\ \| \qquad \qquad \diagdown H \\ CH \text{---} CH \text{------} H \end{array} \right)^{\ddagger} \rightarrow \begin{array}{c} CH_2 \\ \diagup \quad \diagdown \\ CH \qquad CH \\ \| \qquad \quad \| \\ CH \text{------} CH \end{array} + H_2$$

(Planar)

What are the corresponding values for the ratio of the symmetry numbers of reactants and activated complex, ($\sigma_A \times \sigma_B/\sigma^{\ddagger}$ or $\sigma_A/\sigma^{\ddagger}$)?

17. The rate constant of the reaction $Br + H_2 \rightarrow HBr + H$ is $_2k = 10^{13.9}$ $\exp(-18,400/RT)$ cc mole^{-1} sec^{-1}. Calculate the standard energy, enthalpy, entropy and free energy of activation at 500°K taking the standard state as being pure gas at unit concentration (one mole cc^{-1}).

18. Use T.S.T. to calculate the rate constant for $H + H_2 \rightarrow H_2 + H$ at 1000°K. For the hydrogen molecule $d(H–H) = 0\cdot74$Å and $v(H–H) = 4,415\cdot6$ cm^{-1}. Use the BEBO method and the data in question 15 to obtain the height of the potential energy barrier (and hence $E_0^{\ddagger}$) and the bond length in the complex. The $(H_3)^{\ddagger}$ complex has a symmetric stretching vibration of $v = 2,108$ cm^{-1} and two degenerate bending vibrations of $v = 877$ cm^{-1}.

CHAPTER FOUR

COMPLEX PROCESSES: CHAIN REACTIONS

4.1 INTRODUCTION

Having considered in Chapter 3 some of the theories by which the rates of elementary processes can in principle be calculated, we now turn our attention to the question of how the kinetics of overall, observable, chemical changes are determined by the elementary reactions of which they are compounded, that is by their mechanism. We shall consider a few examples of the more well-known chemically complex gas-phase processes, especially the types of processes known as chain reactions. These chain processes are important not only because of their widespread distribution throughout chemistry and in many practically important fields such as pyrolysis and combustion, but also because the experimental character-istics of their kinetics are so vastly different from the behaviour to be expected for elementary reactions and their more simple combinations. In this chapter we seek the connexions between these characteristics and the reaction mechanisms. But first we will summarize briefly the principal types of elementary reactions from which these complex processes are built up.

4.2 TYPES OF ELEMENTARY REACTIONS

Elementary homogeneous gas-phase reactions can be conveniently classi-fied according to their molecularity.

4.2a Unimolecular reactions

These are of two main types, isomerizations and dissociations. Isomeriza-tion reactions are of the general type,

$$A^* \rightarrow B \tag{4.1}$$

and may be due to structural geometrical or optical rearrangement. Since only one product molecule is formed, all the internal energy of A together with any energy released in the isomerization must reside as internal energy of B immediately after the isomerization step since conservation of momentum prohibits any transfer of energy into translational motion of the product B. Until B loses this internal energy by transfer to other mole-cules in collisions it remains capable of re-isomerizing back to A. In dissociation reactions, however, two or possibly three fragments are formed,

$$A^* \rightarrow B + C \qquad [4.2]$$

and so energy can appear as relative translational motion of the products. The most common type of dissociation reaction of normal molecules in the gas-phase leads to the formation of two free radicals but some molecular eliminations in which two stable molecules are formed are also known. Examples of all these types of unimolecular reactions are given in Table 4.1.

TABLE 4.1 Examples of unimolecular reactions*

Reaction	$\log_{10}k$ (sec^{-1}) = $\log_{10}A - E_a/\theta$; θ = $2 \cdot 3RT$ (kcal mole^{-1})
Isomerizations	
cyclopropane $\rightarrow$ propylene	$15 \cdot 45 - 65 \cdot 6/\theta$
cis-1,2-dideuteriocyclopropane $\rightarrow$ *trans*-1,2-dideuteriocyclopropane	$16 \cdot 4 - 65 \cdot 1/\theta$
cis-1,2-dimethylcyclopropane $\rightarrow$ *trans*	$15 \cdot 25 - 59 \cdot 42/\theta$
,, $\rightarrow$ 2-methylbutene-1	$13 \cdot 93 - 61 \cdot 9/\theta$
,, $\rightarrow$ 2-methylbutene-2	$14 \cdot 08 - 62 \cdot 3/\theta$
,, $\rightarrow$ *cis*-pentene-2	$13 \cdot 92 - 61 \cdot 4/\theta$
,, $\rightarrow$ *trans*-pentene-2	$13 \cdot 96 - 61 \cdot 2/\theta$
trans-1,2-dimethylcyclopropane	
$\rightarrow$ 2-methylbutene-1	$13 \cdot 93 - 61 \cdot 9/\theta$
,, $\rightarrow$ 2-methylbutene-2	$14 \cdot 08 - 62 \cdot 3/\theta$
,, $\rightarrow$ *cis*-pentene-2	$14 \cdot 40 - 63 \cdot 6/\theta$
,, $\rightarrow$ *trans*-pentene-2	$14 \cdot 30 - 62 \cdot 9/\theta$
$CH_3NC \rightarrow CH_3CN$	$13 \cdot 6 - 38 \cdot 4/\theta$
$\overline{CH_2CH_2}O \rightarrow CH_3CHO$	$14 \cdot 5 - 57 \cdot 0/\theta$
$\overline{CH_2CH_2}C(CH_3)CH{=}CH_2 \rightarrow \overline{CH_2(CH_2)_2CH}{=}CCH_3$	$14 \cdot 11 - 49 \cdot 35/\theta$
Dissociations	
To radicals	
$C_2H_6 \rightarrow 2CH_3^{\cdot}$	$16 \cdot 5 - 88/\theta$
$C_2H_5^{\cdot} \rightarrow C_2H_4 + H$	$13 - 40/\theta$
$n\text{-}C_3H_7^{\cdot} \rightarrow C_3H_6 + H$	$14 \cdot 6 - 38/\theta$
$i\text{-}C_3H_7^{\cdot} \rightarrow C_3H_6 + H$	$14 \cdot 2 - 37 \cdot 5/\theta$
$n\text{-}C_3H_7^{\cdot} \rightarrow C_2H_4 + CH_3^{\cdot}$	$(12 \cdot 2 \pm 3) - (27 \pm 7)/\theta$
$N_2O_5 \rightarrow NO_2 + NO_3$	$(14 \cdot 8 \pm 1 \cdot 5) - (21 \pm 2)/\theta$

TABLE 4.1 (*continued*)

$N_2O \rightarrow N_2 + O$	$11 \cdot 9$–$60/\theta$
$CH_2N_2 \rightarrow CH_2 + N_2$	$12 \cdot 08$–$34 \cdot 0/\theta$
$CH_3NNCH_3 \rightarrow 2CH_3^* + N_2$	$17 \cdot 24$–$55 \cdot 5/\theta$
$CH_3CO^* \rightarrow CH_3^* + CO$	$15 \cdot 0$–$10 \cdot 3/\theta$
$Pb(C_2H_5)_4 \rightarrow Pb(C_2H_5)_3^* + C_2H^*$	$12 \cdot 6$–$37/\theta$
$CH_3OCH_2^* \rightarrow CH_3^* + CH_2O$	$13 \cdot 2$–$25 \cdot 5/\theta$
$CH_3OC_2H_5 \rightarrow CH_3^* + CH_3OCH_2^*$	$15 \cdot 7$–$78 \cdot 7/\theta$
$CH_3OOCH_3 \rightarrow 2CH_3O^*$	$15 \cdot 4$–$36 \cdot 1/\theta$
$CH_3ONO \rightarrow CH_3O^* + NO$	$13 \cdot 25$–$36 \cdot 4/\theta$
$CH_3NO_2 \rightarrow CH_3^* + NO_2$	$13 \cdot 6$–$50 \cdot 6/\theta$
To molecules	
cyclobutane $\rightarrow 2C_2H_4$	$15 \cdot 6$–$62 \cdot 5/\theta$
perfluorocyclobutane $\rightarrow 2C_2F_4$	$15 \cdot 95$–$74 \cdot 1/\theta$
methylcyclobutane $\rightarrow C_3H_6 + C_2H_4$	$15 \cdot 38$–$61 \cdot 2/\theta$
ethylcyclobutane $\rightarrow$ butene-1 $+ C_2H_4$	$15 \cdot 56$–$62 \cdot 0/\theta$
$C_2H_5I \rightarrow C_2H_4 + HI$	$13 \cdot 36$–$50 \cdot 0/\theta$
$i\text{-}C_3H_7I \rightarrow C_3H_6 + HI$	$12 \cdot 96$–$43 \cdot 5/\theta$
ethylacetate $\rightarrow CH_3COOH + C_2H_4$	$12 \cdot 5$–$47 \cdot 8/\theta$
$HOOCCOOH \rightarrow HCOOH + CO_2$	$11 \cdot 9$–$30 \cdot 0/\theta$
$CH_3COOC(CH_3)_2 \rightarrow CH_3COOH + CH_3CH{=}CH_2$	$13 \cdot 42$–$46 \cdot 34/\theta$
$(CH_3)_2\overline{C{-}N}{=}N \rightarrow CH_3CH{=}CH_2 + N_2$	$13 \cdot 89$–$33 \cdot 17/\theta$

* Where several recent measurements differ the spread of values is indicated.

4.2b Bimolecular reactions

These are by far the most common type of elementary reaction. They may be classified into associations and exchanges. Association reactions are the exact reverses of dissociations,

$$B + C \rightarrow A^* \qquad [4.3]$$

and again A^* has sufficient internal energy to dissociate back to $B + C$ until it is removed, for example, by collision with another gas molecule. Radical combinations are common but molecular addition reactions are rarer as suggested above when considering the reverse reactions. Radical and biradical additions to double bonds are another common type of reaction which fall into this category. Biradical insertion reactions likewise have been postulated, see Table 4.2. Exchange reactions are of the general type,

$$A + B \rightarrow C + D \qquad [4.4]$$

This includes the very common abstraction reactions of free radicals in which an atom, usually a hydrogen atom, or rarely a more complex radical, is transferred from a molecule to the attacking radical. Disproportionation reactions which always accompany the combinations of organic free radicals also fall into this very wide category. Some examples of these bimolecular reactions are given in Table 4.2.

TABLE 4.2 Examples of bimolecular reactions*

Reaction	$\log_{10}k$ $(cc\,mole^{-1}\,sec^{-1})$ $= \log_{10}A - E_a/\theta;$ $\theta = 2 \cdot 3RT\,kcal\,mole^{-1}$
Associations (high pressure)	
Radical + radical	
$CH_3^{\cdot} + CH_3^{\cdot} \rightarrow C_2H_6$	13·3
$C_2H_5^{\cdot} + C_2H_5^{\cdot} \rightarrow C_4H_{10}$	13·2
$CH_3^{\cdot} + C_2H_5^{\cdot} \rightarrow C_3H_8$	13·8
$CF_3^{\cdot} + CF_3^{\cdot} \rightarrow C_2F_6$	13·4
$CCl_3^{\cdot} + CCl_3^{\cdot} \rightarrow C_2Cl_6$	13·5
Radical + molecule	
$H + C_2H_4 \rightarrow C_2H_5^{\cdot}$	$11\cdot0 \pm 1\cdot0$
$H + C_3H_6 \rightarrow C_3H_7^{\cdot}$	$10\cdot7 \pm 0\cdot9$
$Cl + C_2Cl_4 \rightarrow C_2Cl_5^{\cdot}$	$11\cdot4 + 0\cdot5/\theta$
$CH_3^{\cdot} + C_2H_4 \rightarrow C_3H_7^{\cdot}$	$(11\cdot0 \pm 1\cdot0)-$ $(7\cdot0 \pm 1\cdot5)/\theta$
$C_2H_5^{\cdot} + C_2H_4 \rightarrow C_4H_9^{\cdot}$	$(11\cdot0 \pm 1\cdot0)-$ $(7\cdot0 \pm 1\cdot5)/\theta$
$C_2H_5^{\cdot} + 1\text{-hexene} \rightarrow C_8H_{17}^{\cdot}$	$11\cdot2 - 7\cdot8/\theta$
$C_2H_5^{\cdot} + 1\text{-heptene} \rightarrow C_9H_{19}^{\cdot}$	$(11\cdot2 \pm 0\cdot1)-$ $(7\cdot5 \pm 0\cdot5)/\theta$
$C_2H_5^{\cdot} + 1\text{-octene} \rightarrow C_{10}H_{21}^{\cdot}$	$11\cdot1 - 7\cdot7/\theta$
$CH_3^{\cdot} + CH{\equiv}CH \rightarrow CH_3CH{=}CH$	$11\cdot4 - 7\cdot7/\theta$
$C_2H_5^{\cdot} + CH{\equiv}CH \rightarrow CH_3CH_2CH{=}CH$	$11\cdot0 - 7\cdot0/\theta$
$C_2H_5^{\cdot} + 2,3\text{-dimethylbutadiene-1,3} \rightarrow C_8H_{15}^{\cdot}$	$11\cdot5 - 4\cdot5/\theta$
$O(^3P) + C_2H_4 \rightarrow C_2H_4O^*$	$11\cdot3 \pm 0\cdot5$
$CH_2 + H_2 \rightarrow CH_4^*$	?
$CH_2 + C_2H_4 \rightarrow \text{cyclopropane}^*$	?
$CH_2 + CH_4 \rightarrow C_2H_6^*$	?
Molecule + molecule	
$HI + C_2H_4 \rightarrow C_2H_5I$	$11\cdot5 - 28\cdot9/\theta$
$HI + C_3H_6 \rightarrow i\text{-}C_3H_7I$	$10\cdot9 - 23\cdot4/\theta$
$2C_2F_4 \rightarrow \text{perfluorocyclobutane}$	$(10\cdot6 \pm 0\cdot2)-$ $(24\cdot8 \pm 0\cdot8)/\theta$

TABLE 4.2 (*continued*)

Exchanges

H-abstractions

$H + H_2 \rightarrow H_2 + H$	$14 \cdot 0 - 8 \cdot 0/\theta$
$H + D_2 \rightarrow HD + D$	$12 \cdot 6 - 7 \cdot 3/\theta$
$H + CH_4 \rightarrow H_2 + CH_3^{\cdot}$	$(13 \pm 1 \cdot 5) - (10 \pm 4)/\theta$
$H + C_2H_6 \rightarrow H_2 + C_2H_5^{\cdot}$	$14 \cdot 1 - 9 \cdot 7/\theta$
$H + C_3H_8 \rightarrow H_2 + C_3H_7^{\cdot}$	$(13 \cdot 9 \pm 0 \cdot 1) -$
	$(8 \cdot 3 \pm 0 \cdot 5)/\theta$
$H + HF \rightarrow H_2 + F$	$13 \cdot 6 - 33 \cdot 3/\theta$
$H + HCl \rightarrow H_2 + Cl$	$13 \cdot 6 - 4 \cdot 5/\theta$
$H + HBr \rightarrow H_2 + Br$	$13 \cdot 2 - 0 \cdot 9/\theta$
$H + HI \rightarrow H_2 + I$	$12 \cdot 0$
$H + HNO \rightarrow H_2 + NO$	$12 \cdot 8$
$O + H_2 \rightarrow OH^{\cdot} + H$	$13 \cdot 5 - 9 \cdot 5/\theta$
$CH_3^{\cdot} + H_2 \rightarrow CH_4 + H$	$11 \cdot 7 - 10 \cdot 5/\theta$
$CH_3^{\cdot} + CH_4 \rightarrow CH_4 + CH_3^{\cdot}$	$11 \cdot 2 - 12 \cdot 8/\theta$
$CH_3^{\cdot} + C_2H_6 \rightarrow CH_4 + C_2H_5^{\cdot}$	$11 \cdot 5 - 10 \cdot 8/\theta$
$CH_3^{\cdot} + C_3H_8 \rightarrow CH_4 + C_3H_7^{\cdot}$	$11 \cdot 4 - 10 \cdot 1/\theta$
$CH_3^{\cdot} + HI \rightarrow CH_4 + I$	$12 \cdot 5 - 2 \cdot 3/\theta$
$CH_3^{\cdot} + CH_3F \rightarrow CH_4 + CH_2F^{\cdot}$	$11 \cdot 5 - 9 \cdot 1/\theta$
$CH_3^{\cdot} + CHCl_3 \rightarrow CH_4 + CCl_3^{\cdot}$	$10 \cdot 6 - 6 \cdot 2/\theta$
$CH_3^{\cdot} + NH_3 \rightarrow CH_4 + NH_2^{\cdot}$	$11 \cdot 0 - 10 \cdot 9/\theta$
$CH_3^{\cdot} + CH_3CHO \rightarrow CH_4 + CH_3CO^{\cdot}$	$12 \cdot 0 - 7 \cdot 8/\theta$
$CH_3^{\cdot} + CH_3OCH_3 \rightarrow CH_4 + {}^{\cdot}CH_2COCH_3$	$11 \cdot 8 - 9 \cdot 9/\theta$
$CH_3^{\cdot} + CH_3OH \rightarrow CH_4 + {}^{\cdot}CH_2OH$	$(10 \cdot 4 \pm 1 \cdot 0) -$
	$(8 \cdot 4 \pm 2 \cdot 0)/\theta$
$CH_3^{\cdot} + CH_3COCH_3 \rightarrow CH_4 + {}^{\cdot}CH_2COCH_3$	$11 \cdot 9 - 10 \cdot 1/\theta$
$C_2H_5^{\cdot} + H_2 \rightarrow C_2H_6 + H$	$11 \cdot 8 - 11 \cdot 3/\theta$
$C_2H_5^{\cdot} + D_2 \rightarrow C_2H_5D + D$	$12 \cdot 3 - 13 \cdot 3/\theta$
$C_2H_5^{\cdot} + nC_4H_{10} \rightarrow C_2H_6 + C_4H_9^{\cdot}$	$11 \cdot 3 - 10 \cdot 4/\theta$
$C_2H_5^{\cdot} + HI \rightarrow C_2H_6 + I$	$11 \cdot 9 - 1 \cdot 1/\theta$
$C_2H_5^{\cdot} + C_2H_5CHO \rightarrow C_2H_6 + C_2H_5CO^{\cdot}$	$11 \cdot 1 - 5 \cdot 9/\theta$
$C_2H_5^{\cdot} + C_2H_5COC_2H_5 \rightarrow C_2H_6 + {}^{\cdot}C_2H_4COC_2H_5$	$11 \cdot 6 - (8 \cdot 3 \pm 0 \cdot 5)/\theta$
$C_2H_5^{\cdot} + C_2H_5^{\cdot} \rightarrow C_2H_4 + C_2H_6$	$12 \cdot 4$
$F + H_2 \rightarrow HF + H$	$13 \cdot 8 - 1 \cdot 7/\theta$
$F + CH_4 \rightarrow HF + CH_3^{\cdot}$	$13 \cdot 5 - 1 \cdot 2/\theta$
$F + C_2H_6 \rightarrow HF + C_2H_5^{\cdot}$	$13 \cdot 0 - 0 \cdot 3/\theta$
$Cl + H_2 \rightarrow HCl + H$	$13 \cdot 9 - 5 \cdot 5/\theta$
$Cl + CH_4 \rightarrow HCl + CH_3^{\cdot}$	$13 \cdot 4 - 3 \cdot 8/\theta$
$Cl + C_2H_6 \rightarrow HCl + C_2H_5^{\cdot}$	$14 \cdot 0 - 1 \cdot 0/\theta$
$Cl + C_3H_8 \rightarrow HCl + C_3H_7^{\cdot}$	$13 \cdot 7 - 0 \cdot 7/\theta$
$Cl + CH_3Cl \rightarrow HCl + CH_2Cl^{\cdot}$	$13 \cdot 5 - 3 \cdot 3/\theta$
$Cl + CH_2Cl_2 \rightarrow HCl + CHCl_2^{\cdot}$	$13 \cdot 4 - 3 \cdot 0/\theta$
$Cl + CHCl_3 \rightarrow HCl + CCl_3^{\cdot}$	$12 \cdot 8 - 3 \cdot 4/\theta$
$Br + H_2 \rightarrow HBr + H$	$(13 \cdot 9 \pm 0 \cdot 5) -$
	$(18 \cdot 4 \pm 1 \cdot 0)/\theta$
$Br + CH_4 \rightarrow HBr + CH_3^{\cdot}$	$13 \cdot 8 - 18 \cdot 2/\theta$
$Br + C_2H_6 \rightarrow HBr + C_2H_5^{\cdot}$	$13 \cdot 9 - 13 \cdot 2/\theta$
$Br + CH_3OH \rightarrow HBr + CH_2OH^{\cdot}$	$11 \cdot 9 - 6 \cdot 6/\theta$

TABLE 4.2 (*continued*)

$Br + CHF_3 \rightarrow HBr + CF_3^{\cdot}$	$13 \cdot 5 - 23 \cdot 5/\theta$
$Br + CH_3Br \rightarrow HBr + CH_2Br^{\cdot}$	$13 \cdot 2 - 15 \cdot 9/\theta$
$Br + CH_3Cl \rightarrow HBr + CH_2Cl^{\cdot}$	$13 \cdot 6 - 14 \cdot 3/\theta$
$I + H_2 \rightarrow HI + H$	$14 \cdot 1 - 33 \cdot 5/\theta$
$I + CH_4 \rightarrow HI + CH_3^{\cdot}$	$14 \cdot 6 - 32 \cdot 4/\theta$
$I + C_2H_6 \rightarrow HI + C_2H_5^{\cdot}$	$14 \cdot 2 - 27 \cdot 9/\theta$
$CF_3^{\cdot} + CH_4 \rightarrow CF_3H + CH_3^{\cdot}$	$11 \cdot 8 - 10 \cdot 3/\theta$
$CF_3^{\cdot} + C_2H_6 \rightarrow CF_3H + C_2H_5^{\cdot}$	$11 \cdot 7 - 7 \cdot 5/\theta$
$CF_3^{\cdot} + C_3H_8 \rightarrow CF_3H + C_3H_7^{\cdot}$	$11 \cdot 7 - 6 \cdot 2/\theta$
$CH_3O^{\cdot} + CH_4 \rightarrow CH_3OH + CH_3^{\cdot}$	$11 \cdot 8 - 11 \cdot 0/\theta$
$CH_3O^{\cdot} + C_2H_6 \rightarrow CH_3OH + C_2H_5^{\cdot}$	$11 \cdot 4 - 7 \cdot 1/\theta$
$CH_3O^{\cdot} + C_3H_8 \rightarrow CH_3OH + C_3H_7^{\cdot}$	$11 \cdot 2 - 5 \cdot 2/\theta$

Other abstractions

$H + O_2 \rightarrow OH^{\cdot} + O$	$(14 \cdot 5 \pm 0 \cdot 2) - (17 \cdot 0 \pm 0 \cdot 5)/\theta$
$H + O_3 \rightarrow OH^{\cdot} + O_2$	$13 \cdot 2$
$H + NO_2 \rightarrow OH^{\cdot} + NO$	$13 \cdot 5$
$O + OH^{\cdot} \rightarrow H + O_2$	$13 \cdot 5$
$O + NO_2 \rightarrow O_2 + NO$	$13 \cdot 3 - 1 \cdot 1/\theta$
$N + O_2 \rightarrow NO + O$	$7 \cdot 7$
$N + O_3 \rightarrow NO + O_2$	$11 \cdot 5$
$N + NO \rightarrow N_2 + O$	$13 \cdot 1$
$H + Cl_2 \rightarrow HCl + Cl$	$14 \cdot 5 - 3 \cdot 0/\theta$
$H + Br_2 \rightarrow HBr + Br$	$14 \cdot 2 - 0 \cdot 9/\theta$
$H + I_2 \rightarrow HI + I$	$13 \cdot 2$
$CH_3^{\cdot} + CCl_4 \rightarrow CH_3Cl + CCl_3^{\cdot}$	$13 \cdot 2 - 13 \cdot 4/\theta$
$CH_3^{\cdot} + C_2Cl_6 \rightarrow CH_3Cl + C_2Cl_5^{\cdot}$	$11 \cdot 4 - 10 \cdot 2/\theta$
$Br + HBr \rightarrow Br_2 + H$	$13 \cdot 4 - 41 \cdot 8/\theta$
$Br + CCl_3Br \rightarrow Br_2 + CCl_3^{\cdot}$	$13 \cdot 9 - 10 \cdot 3/\theta$
$Cl + NOCl \rightarrow Cl_2 + NO$	$13 \cdot 1 - 1 \cdot 1/\theta$
$Cl + COCl_2 \rightarrow Cl_2 + COCl^{\cdot}$	$14 \cdot 8 - 20 \cdot 5/\theta$
$Cl + COCl \rightarrow Cl_2 + CO$	$14 \cdot 4 - 0 \cdot 5/\theta$
$I + CH_3I \rightarrow I_2 + CH_3^{\cdot}$	$(14 \cdot 0 \pm 0 \cdot 3) - (19 \cdot 8 \pm 0 \cdot 6)/\theta$
$I + C_2H_5I \rightarrow I_2 + C_2H_5^{\cdot}$	$14 \cdot 0 - 17 \cdot 1/\theta$
$O_3 + NO \rightarrow O_2 + NO_2$	$12 \cdot 0 - 2 \cdot 4/\theta$
$NO_2 + F_2O \rightarrow NO_2F + FO^{\cdot}$	$11 \cdot 1 - 14 \cdot 5/\theta$
$NO_2 + NOCl \rightarrow NO_2Cl + NO$	$10 \cdot 3 - 10 \cdot 0/\theta$
$NO_2 + ClO_2 \rightarrow NO_3 + ClO^{\cdot}$	$10 \cdot 4 - 11 \cdot 0/\theta$
$NO_2 + Cl_2O \rightarrow NO_2Cl + ClO^{\cdot}$	$9 \cdot 8 - 10 \cdot 5/\theta$

Other exchanges (?)[†]

$2NO_2 \rightarrow 2NO + O_2$	$12 \cdot 6 - (25 \cdot 3 \pm 1 \cdot 6)/\theta$
$2NOCl \rightarrow 2NO + Cl_2$	$12 \cdot 7 - 23 \cdot 6/\theta$
$2ClO \rightarrow Cl_2 + O_2$	$(10 \cdot 6 \pm 0 \cdot 5) + (0 \cdot 35 \pm 0 \cdot 65)/\theta$

* Where several recent measurements differ the spread of values is indicated.
† None of these reactions proceeds as a single (4-centre) step.

4.2c Termolecular reactions

As outlined in section 3.3 genuine termolecular gas-phase reactions may not exist. Reactions which are third order are discussed in that section and are of two types. Firstly the very common atom and small radical combinations which require the presence of a chemically inert third body,

$$B + C + M \rightarrow A + M \qquad\qquad [4.5]$$

and secondly the very rare third-order reactions in which all the reactants are involved chemically, such as the reactions of nitric oxide with oxygen, bromine, and chlorine. Some examples of these are summarized in Table 4.3.

TABLE 4.3 Examples of third-order reactions*

Reaction	$\log_{10}k$ (cc^2 mole^{-2} sec^{-1}) = $\log_{10}A - E_a/\theta$, $\theta = 2\cdot3RT$ (kcal mole^{-1})	
Associations:		
$H + H + H_2 \rightarrow 2H_2$	$16\cdot0 \pm 0\cdot1$	$(T = 300°K)$
$O + O + O_2 \rightarrow 2O_2$	$15\cdot0 \pm 0\cdot1$	$(T = 300°K)$
$N + N + N_2 \rightarrow 2N_2$	$15\cdot2 \pm 0\cdot3$	$(T = 300°K)$
$N + O + N_2 \rightarrow NO + N_2$	$15\cdot5$	$(T = 350°K)$
$Cl + Cl + Cl_2 \rightarrow 2Cl_2$	$16\cdot0 \pm 1\cdot5$	$(T = 300°K)$
$Br + Br + Ar \rightarrow Br_2 + Ar$	$15\cdot4$	$(T = 300°K)$
$I + I + Ne \rightarrow I_2 + Ne$	$14\cdot3 + 1\cdot5/\theta$	$(T = 333°K)$
$I + I + I_2 \rightarrow 2I_2$	$14\cdot2 + 5\cdot2/\theta$	
$H + O_2 + Ar \rightarrow HO_2^\cdot + Ar$	$14\cdot67 + 1\cdot6/\theta$	
$O + O_2 + Ar \rightarrow O_3 + Ar$	$12\cdot5 + 2\cdot3/\theta$	
$O + O_2 + N_2 \rightarrow O_3 + N_2$	$13\cdot0 + 1\cdot7/\theta$	
$H + NO + H_2 \rightarrow HNO + H_2$	$16\cdot0 + 0\cdot6/\theta$	
$O + NO + N_2 \rightarrow NO_2 + N_2$	$15\cdot2 + 1\cdot9/\theta$	
$O + NO + O_2 \rightarrow NO_2 + O_2$	$14\cdot9 + 1\cdot8/\theta$	
$CH_3^\cdot + NO + CH_3COCH_3 \rightarrow$ $CH_3NO + CH_3COCH_3$	$17\cdot5$	$(T = 473°K)$
$CH_3^\cdot + O_2 + M \rightarrow CH_3O_2^\cdot + M$	?	
Others:		
$2NO + O_2 \rightarrow 2NO_2$	$9\cdot0 + 1\cdot1/\theta$	
$2NO + Cl_2 \rightarrow 2NOCl$	$9\cdot7 - 3\cdot7/\theta$	
$2NO + Br_2 \rightarrow 2NOBr$	$9\cdot5$	
$2I + H_2 \rightarrow 2HI$	$13\cdot8 - 5\cdot3/\theta$	

* Where several recent measurements differ the spread of values is indicated.

We now wish to see how these relatively simple processes can combine together in various ways to give rise to complex reactions.

4.3　STRAIGHT CHAIN REACTIONS

Many reactions of atoms and free radicals are of the following type,

$$R_1^. + M_1 \rightarrow M_2 + R_2^. \qquad [4.6]$$

where $R_1^.$ and $R_2^.$ are free radicals, which may be the same or different, and M_1 and M_2 are molecules of reactant and product respectively. Because the radical $R_1^.$ contains an odd number of electrons, one of which is therefore unpaired, while M_1 has an even number of electrons, it follows that at least one product species must have an odd number of electrons and be a free radical, in this case $R_2^.$. If the radical $R_2^.$ can undergo a similar reaction to that of $R_1^.$ in which another reactant molecule is converted to a product molecule and another free radical is generated, which can repeat the process and so on, until eventually $R_1^.$ is regenerated and the cycle started again, we have the possibility of an *indefinite sequence* or *chain* of collisions which convert reactant into product being initiated by the original single $R_1^.$ radical. The reaction [4.6] is called a *propogation step* of a straight chain reaction. If no other processes occurred the introduction of a single *active centre* or *chain centre*, as $R_1^.$ is called, into the reactant gas by an *initiation step*, for example, the dissociation of a reactant molecule into two free radicals,

$$M_1 \rightarrow R_1^. + R_3^. \qquad [4.7]$$

could result in the complete transformation of reactant into products by repeated cycles of the chain. *Termination steps* however may destroy the active centres and so limit the average number of cycles, the average *chain length*, to a finite value. In the absence of chain termination the chain would cycle indefinitely and the chain length would be infinite. It is this fact which distinguishes true chain processes from other sequences of consecutive reactions which involve a finite number of steps and no cycles, for example,

$$A \rightarrow 2B^. \qquad [4.8]$$

$$B^. + A \rightarrow C + D^. \qquad [4.9]$$

$$D^. + B^. \rightarrow E \qquad [4.10]$$

where C and E are stable products and $B^.$ and $D^.$ are reactive intermediates. This is not a chain reaction since only a finite number of reactive collisions, two in this case, can follow the initial step. It is simply a set of consecutive reactions involving two reactive intermediates which might be free radicals for example.

Common termination steps of free radical chain reactions are of the following types,

(i) Radical combination,

$$R_1^{\cdot} + R_2^{\cdot} \rightarrow M_3 \qquad\qquad [4.11]$$

in which a stable molecule with all its electrons paired is formed from two active centres. If the radicals involved are small and therefore have few internal degrees of freedom, this combination may need a third body to remove excess energy and stabilize the product (p. 96), thus,

$$R_1^{\cdot} + R_2^{\cdot} + M \rightarrow M_3 + M \qquad\qquad [4.12]$$

This type of termination is called a quadratic termination step since its rate depends on the *square* of an active centre concentration, or the product of two such concentrations if $R_1^{\cdot}$ and $R_2^{\cdot}$ in [4.11] or [4.12] are different radicals.

(ii) Formation of a radical of insufficient reactivity to propogate the chain, for example by combination with a molecule,

$$R_1^{\cdot} + M_4^{\cdot} \rightarrow R_{inactive}^{\cdot} \qquad\qquad [4.13]$$

or by exchange,

$$R_1^{\cdot} + M_5 \rightarrow R_{inactive}^{\cdot} + M_6 \qquad\qquad [4.14]$$

For example $R_{inactive}^{\cdot}$ may be rendered unreactive by virtue of resonance stabilization as occurs in the case of the allyl radical formed by abstraction of a hydrogen atom from propylene. Again the combinative type of reaction [4.13] may need a third body if the species involved are small,

$$R_1^{\cdot} + M_4 + M \rightarrow R_{inactive}^{\cdot} + M \qquad\qquad [4.15]$$

Inactive radicals may also be involved in termination as follows,

$$R_1^{\cdot} + R_{inactive}^{\cdot} \rightarrow M_7 \qquad\qquad [4.16]$$

or

$$R_1^{\cdot} + R_{inactive}^{\cdot} + M \rightarrow M_7 + M \qquad\qquad [4.17]$$

Although these reactions [4.16] and [4.17] are radical combinations similar to those considered in (i), i.e. [4.11] and [4.12], they differ from these in the important respect that the rate depends on the concentration of *active* centres to the first power only as also do those of [4.13], [4.14] and [4.15]. These reactions are all accordingly called linear gas-phase terminations to distinguish them from the quadratic ones, [4.11] and [4.12].

(iii) The last common type of termination involves interaction of the active centres with the surface of the reaction vessel to form an adsorbed species which is subsequently destroyed without further generation of active centres,

$$R_1^\cdot + S \rightarrow R_1 S \qquad [4.18]$$

where S represents an adsorption site on the surface. The rate of this type of chain termination depends on the reaction vessel geometry, its surface to volume ratio, the concentration gradient of active centres at the wall, and the diffusion coefficient of the active centres, since for most radical adsorption processes diffusion to the surface rather than adsorption or subsequent reaction is the slowest step in the overall process of wall termination. As a crude approximation the rate of [4.18] can often be taken as being proportional to a spatially-averaged gas-phase concentration of active centres and hence it may be classed as a linear termination.

4.3a The approach to the steady state and the induction period

(i) *Linearly terminated chains*

Before considering a few specific examples of straight chain mechanisms we will investigate the kinetics of a simplified hypothetical straight chain reaction with the following mechanism,

$$M_1 \xrightarrow{r_i} R^\cdot + R_{\text{inactive}}^\cdot \qquad \text{initiation} \qquad [4.19]$$

$$R^\cdot + M_2 \xrightarrow{r_p} M_3 + R^\cdot \qquad \text{propagation} \qquad [4.20]$$

$$R^\cdot \xrightarrow{r_t} \text{end} \qquad \text{termination} \qquad [4.21]$$

This represents a straight chain propagated by a single chain carrier $R^\cdot$ and terminated linearly. Let the rate per unit volume and the rate constant be denoted by r and k respectively with suffixes i, p and t to denote initiation, propagation and termination reactions, respectively. The rate of change of concentration of active centres in a closed constant-volume system is given by,

$$\frac{d[R^\cdot]}{dt} = r_i - k_t[R^\cdot] \qquad (4.1)$$

since active centres are formed by initiation and destroyed by termination, but are *unaffected* by propagation. The solution of (4.1) subject to the initial condition that $[R^\cdot] = 0$ when $t = 0$ is,

$$[R^\cdot] = \frac{r_i}{k_t}(1 - e^{-k_t t}) \qquad (4.2)$$

as may be readily verified by differentiation. Now the rate of reaction, that is the rate at which M_2 is converted to M_3, is

$$\frac{d[M_3]}{dt} = -\frac{d[M_2]}{dt} = k_p[R^{\cdot}][M_2]$$

$$= k_p \frac{r_i}{k_t}[M_2](1 - e^{-k_t t}) \tag{4.3}$$

or

$$-\frac{d\ln[M_2]}{dt} = k_p \frac{r_i}{k_t}(1 - e^{-k_t t}) \tag{4.4}$$

The amount of reaction that has occurred after a time t is found by integrating (4.4) with the initial condition that $[M_2] = [M_2]_0$ at $t = 0$. Thus,

$$\ln \frac{[M_2]_0}{[M_2]} = \frac{r_i k_p}{k_t}\left\{t + \frac{1}{k_t}(e^{-k_t t} - 1)\right\} \tag{4.5}$$

Figure 4.1 shows the concentration of active centres given by equation (4.2) and similarly Figure 4.2 shows the amount of reaction as given by

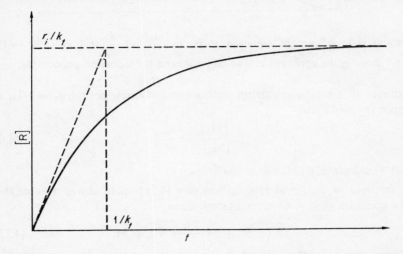

FIGURE 4.1   A plot of the radical concentration $[R^{\cdot}] = \dfrac{r_i}{k_t}(1 - e^{-k_t t})$, from equation (4.2) versus time t, showing the approach to a steady state and the induction period $1/k_t$.

(4.5). It is seen that the reaction has an *induction period*, given by k_t^{-1}, and that after a time which is long compared to this the reaction proceeds negligibly far displaced from a steady state as defined in section 1.6. The

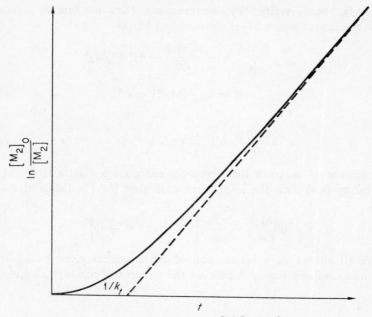

FIGURE 4.2   A plot of equation (4.5), $\ln\dfrac{[M_2]_0}{[M_2]} = \dfrac{r_i k_p}{k_t}\{t + k_t^{-1}(e^{-k_t t} - 1)\}$
showing the approach to a steady state and the induction period $1/k_t$.

amount of reaction occurring during the induction period, $t_i = 1/k_t$, is
given from (4.5) as,

$$\ln\frac{[M_2]_0}{[M_2]_{t_i}} = \frac{r_i k_p}{e k_t^2} \tag{4.6}$$

(ii) *Quadratically terminated chains*

If in place of the linear termination step [4.21] taken above we consider
the common case of quadratic termination,

$$R^{\cdot} + R^{\cdot} + M \xrightarrow{\;r_t\;} M_4 + M \tag{4.22}$$

the equation for the rate of change of the active centre concentration
becomes,

$$\frac{d[R^{\cdot}]}{dt} = r_i - 2k_t[R^{\cdot}]^2[M] \tag{4.7}$$

Putting,

$$x = \left(\frac{2k_t[M]}{r_i}\right)^{\frac{1}{2}}[R^{\cdot}] \tag{4.8}$$

this integrates as follows,

$$\left(\frac{2k_t[\text{M}]}{r_i}\right)^{\frac{1}{2}} t = \frac{1}{r_i} \int_0^x \frac{\mathrm{d}x}{1 - x^2} \tag{4.9}$$

or,

$$(2r_ik_t[\text{M}])^{\frac{1}{2}}t = \frac{1}{2} \ln \left(\frac{1 + x}{1 - x}\right) \tag{4.10}$$

At very long times again a steady state is reached in which, from (4.10)

$$x = 1$$

i.e.

$$[\text{R}^\cdot]_{s.s.} = \left(\frac{r_i}{2k_t[\text{M}]}\right)^{\frac{1}{2}} \tag{4.11}$$

This is the same result that is obtained by setting $\mathrm{d}[\text{R}^\cdot]/\mathrm{d}t = 0$ in (4.7) following the usual steady state method. The induction period t_x, or time taken to reach a fraction x of the steady state concentration of active centres $[\text{R}^\cdot]_{s.s.}$ is given from (4.10) as,

$$t_x = \frac{1}{2(2r_ik_t[\text{M}])^{\frac{1}{2}}} \ln \left(\frac{1 + x}{1 - x}\right) \tag{4.12}$$

For example to reach within 10% of the steady state concentration where $x = 0\cdot9$

$$t_{0\cdot9} \simeq \frac{3}{2(2r_ik_t[\text{M}])^{\frac{1}{2}}} \tag{4.13}$$

Equation (4.12) may be written,

$$t_x = \tau \tanh^{-1}x \tag{4.14}$$

where,

$$\tau = \frac{1}{(2r_ik_t[\text{M}])^{\frac{1}{2}}} \tag{4.15}$$

From (4.15) and (4.11),

$$r_i\tau = [\text{R}^\cdot]_{s.s.} \tag{4.16}$$

hence τ is seen to be the average life of a chain centre under steady state conditions (see footnote p. 135). From (4.14),

$$x = \tanh (t_x/\tau) \tag{4.17}$$

The rate of reaction is given as before by,

$$-\frac{\mathrm{d}[\text{M}_2]}{\mathrm{d}t} = k_p[\text{R}^\cdot][\text{M}_2]$$

Making the substitution (4.8),

$$-\frac{d\ln [M_2]}{dt} = k_p \left(\frac{r_i}{2k_t[M]}\right)^{\frac{1}{2}} x$$

which from (4.11), (4.16) and (4.17) is,

$$-\frac{d\ln [M_2]}{dt} = k_p r_i \tau \tanh (t_x/\tau) \tag{4.18}$$

Integration between $t_x = 0$ and t gives,

$$\ln \frac{[M_2]_0}{[M_2]} = k_p r_i \tau^2 \ln [\cosh(t/\tau)] \tag{4.19}$$

or,

$$\frac{[M_2]_0}{[M_2]} = [\cosh (t/\tau)]^{k_p/2k_t[M]} \tag{4.20}$$

Substituting $[1 - \tanh^2(t/\tau)]^{-\frac{1}{2}}$ for $\cosh (t/\tau)$ and using (4.17) we get,

$$\ln \frac{[M_2]_0}{[M_2]} = -\frac{k_p}{4k_t[M]} \ln (1 - x^2) \tag{4.21}$$

This equation enables us to calculate the amount of reaction that occurs during the induction period. If, for example, not more than 10% reaction is to occur during the approach to within 10% of the steady state, substitution of $[M_2]_0/[M_2] \leqslant 100/90$ and $x = 0.9$ into (4.21) gives

$$\frac{k_p}{4k_t[M]} \leqslant 0.06 \tag{4.22}$$

This result thus gives us a criterion for judging whether or not most of the reaction occurs in the steady state, which will be useful when we come to study some specific examples of quadratically terminated straight chains.

Although the discussion in this section has been limited so far to a particularly simple type of chain mechanism which has only a single type of chain centre, it may be extended to more complex cases by employing the quasi-stationary state method outlined in section 1.6c. Thus in a chain reaction involving many different chain centres there is an initial period of reaction during which the *relative* concentrations of these active centres reach steady values. This is often a very short time, and after this the reaction can be characterized by means of a single concentration variable for all the active centres, as occurs in the special case of a one-centre chain considered above, since,

$$[R_1^\cdot] \propto [R_2^\cdot] \propto [R_3^\cdot] \ldots \propto \Sigma[R_n] \qquad (4.23)$$

The approach of the total active centre concentration towards the steady state value may then be described by applying the methods of sections (i) or (ii) above to this quasi-stationary state. Thus many chain reactions have measurable induction or incubation periods and many do in fact proceed mainly in stationary or quasi-stationary states. Let us now consider a few simple examples of straight chain reactions.

4.3b The thermal conversion of para to ortho hydrogen

Considering for simplicity the reaction of pure para hydrogen so that the reverse process can be neglected, this reaction proceeds as follows,

$$H_2 + S \xrightarrow{k_i} 2H + S \qquad \text{initiation} \qquad [4.23]$$

$$H + p\text{-}H_2 \xrightarrow{k_p} o\text{-}H_2 + H \qquad \text{propagation} \qquad [4.24]$$

$$2H + S \xrightarrow{k_t} H_2 + S \qquad \text{termination} \qquad [4.25]$$

where S represents the surface of the reaction vessel which catalyses the dissociation and recombination of hydrogen. This is an example of a straight chain with one chain carrier, hydrogen atoms, and mutual, quadratic, termination of chains. Note again that the propagation reaction has no effect on the concentration of active centres [H] since one hydrogen atom is formed for every one that reacts converting para to ortho hydrogen.

The induction periods observed experimentally are of about 2 to 3 minutes duration under typical conditions. This shows that initiation and termination cannot be the homogeneous processes,

$$H_2 + M \underset{k_t'}{\overset{k_i'}{\rightleftarrows}} 2H + M \qquad [4.26]$$

since from equation (4.13) the predicted induction period is,

$$t_{0\cdot 9} = \frac{3}{2(2r_i'k_t'[M])^{\frac{1}{2}}}$$

where,

$$r_i' = 2k_i'[\text{H}_2][\text{M}]$$

so that

$$t_{0.9} = \frac{3}{4(k_i'k_t'[\text{H}_2])^{\frac{1}{2}}[\text{M}]}$$

Taking values for the homogeneous rate constants given by shock tube measurements of hydrogen dissociation, $k_i' = 5 \times 10^{15} \exp(-103{,}000/RT)$ cc mol^{-1} sec^{-1} and $k_t' = 5 \times 10^{15}$ cc^2 mole^{-2} sec^{-1}, together with the typical conditions of 1000°K and 10 torr pressure we get $t_{0.9} = 5 \times 10^5$ sec. Since experimentally it is found that the steady state is reached far more rapidly than this it must be concluded that the initiation and termination rates do not correspond to those for the homogeneous gas-phase reactions [4.26] but must be considerably faster. Acceleration of both forward and reverse steps by about a factor of 5×10^3 is needed to account for the observed value of the induction period. Catalysis of these steps by the surface of the reaction vessel accounts for the fast approach to steady state conditions that is observed in practice.

After sufficient time therefore the reaction will be negligibly displaced from a steady state in which initiation and termination are balanced; in this case actually an equilibrium since initiation and termination are the reverses of each other. Thus,

$$[\text{H}]^2 = \frac{k_i}{k_t}[\text{H}_2] \qquad\qquad (4.24)$$

$$= K[\text{H}_2] \qquad\qquad (4.25)$$

where K is the dissociation constant of hydrogen (see section 1.5b). The rate of conversion of para to ortho hydrogen in the propagating step [4.24] is,

$$\frac{d[\text{o-H}_2]}{dt} = k_p[\text{H}][\text{p-H}_2] \qquad\qquad (4.26)$$

Provided the rate of termination of the chain by [4.25] (which will produce some ortho hydrogen) can be neglected in comparison to this propagation rate (see below) this latter will represent the total rate of formation of ortho hydrogen from para. Substitution for the steady state hydrogen atom concentration from (4.24) into (4.26) gives,

$$\frac{d[\text{o-H}_2]}{dt} = k_p \left(\frac{k_i}{k_t}\right)^{\frac{1}{2}} [\text{H}_2]^{\frac{1}{2}}[\text{p-H}_2] \qquad\qquad (4.27)$$

which for pure para hydrogen becomes,

$$\frac{d[o\text{-}H_2]}{dt} = k_p \left(\frac{k_i}{k_t}\right)^{\frac{1}{2}} [p\text{-}H_2]^{\frac{3}{2}} \qquad (4.28)$$

The steady state reaction is therefore predicted to be of $\frac{3}{2}$-order as is found to be the case experimentally. Half-integral orders such as this occur not infrequently in chain reactions involving quadratic termination as may be expected from the general form of the steady state equations given above.

Whether it is legitimate to ignore the rate of formation of ortho hydrogen in the termination step, i.e.

$$\frac{d[o\text{-}H_2]}{dt} = \tfrac{3}{4}k_t[H]^2[S] = \tfrac{3}{4}k_i[H_2][S] \qquad (4.29)$$

in comparison with the propagation rate (4.26) depends on the chain length. This is the average number of chain cycles completed by a single hydrogen atom,

$$\text{chain length, } n = \frac{\text{rate of propagation}}{\text{rate of initiation or termination}} \qquad (4.30)$$

which from (4.28) is,

$$n = \frac{k_p}{2(k_i k_t)^{\frac{1}{2}}} \frac{[p\text{-}H_2]^{\frac{1}{2}}}{[S]} \qquad (4.31)$$

From equations (4.13) and (4.15) it is seen that the average life of a radical chain, τ, is of the same order of magnitude as the induction period, e.g. $t_{0.9}$, and τ is related directly to n since from (4.16),

$$\tau = \frac{[H]_{\text{s.s.}}}{r_i}$$

and from (4.30),

$$n = \frac{k_p[H]_{\text{s.s.}}.[p\text{-}H_2]}{r_i}$$

therefore,

$$n = k_p[p\text{-}H_2]\tau \qquad (4.32)$$

Taking the experimental value for $t_{0.9}$ as about 10^2 sec and the value of $k_p = 10^{14.0} \exp(-8,000/RT)$ cc mole^{-1} sec^{-1}, n is found from (4.32) to be about 10^7 under the typical conditions, $T = 1000°K$, $P = 10$ torr. This is therefore a very long chain reaction and the total rate of reaction is equal to the rate of chain propagation to a very good approximation.

Another feature of chain reactions illustrated by this example appears if we use (4.25) to calculate the actual concentration of active centres that propagate the chain in the steady state. Taking the same typical conditions $T = 1000°K, P = 10$ torr, the partial pressure of hydrogen atoms is found to be about 10^{-7} torr or about 1 part in 10^8 of the total gas. Such exceedingly low concentrations of active centres are common for chains propagated by such reactive species. Another common feature of such chains is illustrated by the overall activation energy for this reaction. This is defined by equation (1.47),

$$E_a = - \frac{\mathrm{dln}_{\frac{3}{2}}k}{\mathrm{d}(1/RT)}$$

where $_{\frac{3}{2}}k$ is the $\frac{3}{2}$-order rate constant of (4.28), i.e.

$$_{\frac{3}{2}}k = k_p \left(\frac{k_i}{k_t}\right)^{\frac{1}{2}} \tag{4.33}$$

or,

$$\ln_{\frac{3}{2}}k = \ln k_p + \tfrac{1}{2}\ln k_i - \tfrac{1}{2}\ln k_t$$

Differentiation with respect to $(1/RT)$ gives,

$$E_a = E_p + \tfrac{1}{2}(E_i - E_t) \tag{4.34}$$

where the activation energies of the various individual elementary rate constants are denoted by the same suffixes as the rate constants. This activation energy E_a is considerably less than that of the production of active centres E_i. This latter refers to a heterogeneous process and its value is uncertain but the difference between E_i and E_t is accurately known, being the bond dissociation energy of hydrogen. Taking the experimental values, $E_i - E_t = D(\text{H–H}) - RT = 103\text{-}2$ kcal mole^{-1}, $E_p = 8$ kcal mole^{-1} we get,

$$E_a = 8 + \tfrac{1}{2} \times 101 = 58\cdot5$$

in good agreement with the experimental value of 58 kcal mole^{-1}. It is seen that this low overall activation energy is again a consequence of the quadratic nature of the termination reaction which leads to the factor of $\tfrac{1}{2}$ in (4.34). For linearly terminated chain reactions equation (4.3) shows that there is no similar reduction in activation energy to significantly below that of initiation since (4.3) leads to,

$$E_a = E_p + E_i - E_t \tag{4.35}$$

for the overall activation energy of a straight linearly terminated one-carrier chain. For most thermally initiated radical chains, E_t and E_p are small in comparison with E_i, as a consequence of the high reactivity of the active centres and the therefore energetically difficult initial formation of them.

The pre-exponential factor of the $\frac{3}{2}$-order constant for the para–ortho hydrogen conversion is similarly given from (4.33) by,

$$_{\frac{3}{2}}A = A_p \left(\frac{A_i}{A_t}\right)^{\frac{1}{2}} \tag{4.36}$$

again using the suffixes as before. The experimental values of these pre-exponential factors are approximately $A_p \simeq 10^{14}$ cc mole^{-1} sec^{-1}, a fairly typical value for such a simple bimolecular exchange reaction (see p. 92) and $A_i/A_t \simeq 1 \cdot 0$ mole cc^{-1}. Hence, $_{\frac{3}{2}}A = 10^{14}$ cc$^{\frac{1}{2}}$ mole$^{-\frac{1}{2}}$ sec^{-1}. If the reaction were mistakenly taken to be second order then the second-order rate constant,

$$_2k = {_{\frac{3}{2}}k}[\mathrm{H_2}]^{-\frac{1}{2}} \tag{4.37}$$

would have a pre-exponential factor of about $_2A = {_{\frac{3}{2}}A}[\mathrm{H_2}]^{-\frac{1}{2}} = 2 \cdot 5 \times 10^{17}$ cc mole^{-1} sec^{-1} under the typical conditions taken above. This is much higher than the collision frequency Z (see sections 3.2 and 3.10). The corresponding first-order rate constant,

$$_1k = {_{\frac{3}{2}}k}[\mathrm{H_2}]^{\frac{1}{2}} \tag{4.38}$$

would similarly have a pre-exponential factor of about $_1A = {_{\frac{3}{2}}A}[\mathrm{H_2}]^{\frac{1}{2}} = 4 \times 10^{10}$ sec^{-1}. This is much lower than a typical vibration frequency or (kT/h), (see sections 3.6 and 3.12). These values illustrate how the mistaken identification of a complex chain reaction as a simple elementary unimolecular or bimolecular step can lead to some unusual pre-exponential factors which are very different from expectations based on the theories of these elementary reactions outlined in Chapter 3. Such mistakes are not uncommon in the chemical literature.

4.3c The hydrogen–bromine reaction

Having considered a chain reaction propagated by a single active species let us now look at the slightly more complex case of a straight chain involving two active centres. The best-known example of this type of reaction is undoubtedly the reaction between hydrogen and bromine to form hydrogen bromide according to the stoichiometric equation,

$$\mathrm{H_2 + Br_2 = 2HBr} \tag{4.27}$$

When initiated thermally this chain reaction proceeds as follows,

$$Br_2 + M \xrightarrow{\ 1\ } 2Br + M \qquad \text{initiation} \qquad [4.28]$$

$$Br + H_2 \xrightarrow{\ 2\ } HBr + H \qquad \text{propagation} \qquad [4.29]$$

$$H + Br_2 \xrightarrow{\ 3\ } HBr + Br \qquad \text{propagation} \qquad [4.30]$$

$$2Br + M \xrightarrow{\ 4\ } Br_2 + M \qquad \text{termination} \qquad [4.31]$$

If the product HBr is allowed to accumulate in the reaction system it inhibits (i.e. slows down) the reaction owing to the reverse of step (2) occurring thus,

$$H + HBr \xrightarrow{\ 5\ } H_2 + Br \qquad \text{de-propagation} \qquad [4.32]$$

and reconverting product into reactant. Since there are two chain carriers H and Br there are now two propagating steps (2) and (3) in the complete cycle of the chain.

$$Br \underset{3}{\overset{2}{\rightleftharpoons}} H$$

Each chain cycle produces two molecules of product.

When both chain carriers have reached their steady state concentrations we may write,

$$\frac{d[Br]}{dt} = 0 = 2r_1 - r_2 + r_3 - 2r_4 + r_5 \qquad (4.39)$$

$$\frac{d[H]}{dt} = 0 = r_2 - r_3 - r_5 \qquad (4.40)$$

These give us two simultaneous equations from which to calculate the stationary concentrations of the two chain centres. Adding these two equations,

$$0 = r_1 - r_4 \qquad (4.41)$$

i.e.

$$\text{initiation} = \text{termination} \qquad (4.42)$$

This result is a general one for all straight chains since the net result of the propagating steps on the total radical concentration is always zero and therefore in the steady state initiation must exactly balance termination. In this example only one chain carrier, Br, is involved in the initiation and termination steps so that (4.41) gives us directly an expression for the concentration of this centre, thus,

$$k_1[M][Br_2] = k_4[M][Br]^2$$

$$[Br] = \left(\frac{k_1[Br_2]}{k_4}\right)^{\frac{1}{2}} \tag{4.43}$$

Again in this example, as in the para to ortho hydrogen conversion, initiation and termination are mutual reverses so that the steady state concentration of Br is in fact the equilibrium value,

$$[Br] = (K_d[Br_2])^{\frac{1}{2}} \tag{4.44}$$

where K_d is the dissociation constant of bromine.

Substituting (4.43) into (4.40) gives,

$$k_3[H][Br_2] + k_5[H][HBr] = k_2\left(\frac{k_1}{k_4}\right)^{\frac{1}{2}}[Br_2]^{\frac{1}{2}}[H_2]$$

or

$$[H] = \frac{k_2(k_1/k_4)^{\frac{1}{2}}[Br_2]^{\frac{1}{2}}[H_2]}{k_3[Br_2] + k_5[HBr]} \tag{4.45}$$

Now the overall rate of reaction [4.27] is, (see equation 1.12)

$$\frac{1}{2}\frac{d[HBr]}{dt} = \frac{1}{2}(r_2 + r_3 - r_5) \tag{4.46}$$

which from (4.40) is

$$\frac{1}{2}\frac{d[HBr]}{dt} = r_3 \tag{4.47}$$

Substituting (4.45) into (4.47) gives,

$$\frac{d[HBr]}{dt} = \frac{2k_2(k_1/k_4)^{\frac{1}{2}}[Br_2]^{\frac{1}{2}}[H_2]}{1 + (k_5/k_3)[HBr]/[Br_2]} \tag{4.48}$$

In the early* stages of the reaction before the product HBr accumulates this simplifies to,

$$\left(\frac{d[HBr]}{dt}\right)_{t=0} = 2k_2(k_1/k_4)^{\frac{1}{2}}[Br_2]^{\frac{1}{2}}[H_2] \tag{4.49}$$

This may be compared to the very similar result obtained for the one carrier chain (4.27). The order of reaction with respect to bromine is one half and with respect to hydrogen is unity. The overall order is $\frac{3}{2}$. When

* Provided the steady state is reached.

the product HBr accumulates it inhibits the reaction according to (4.48) and this rate law ceases to be of the simple form of equation (1.16) so that the orders of reaction vary with conditions. This rate law may be written,

$$\frac{d[HBr]}{dt} = \frac{A[Br_2]^{\frac{1}{2}}[H_2]}{1 + B[HBr]/[Br_2]} \qquad (4.50)$$

and is exactly of the form that was discovered experimentally by Bodenstein in 1906. The experimental values of A and B are related to the elementary rate constants by,

$$A = 2k_2(k_1/k_4)^{\frac{1}{2}} \qquad (4.51)$$

$$B = k_5/k_3 \qquad (4.52)$$

Since, (see (4.43), (4.44) and section 1.5b),

$$k_1/k_4 = K_d \qquad (4.53)$$

calculation of K_d thermodynamically or by statistical mechanics enables k_2 to be obtained from the experimental value of A,

$$k_2 = \frac{1}{2}A(K_d)^{-\frac{1}{2}} \qquad (4.54)$$

The equilibrium constant for reaction steps (2) and (5) i.e.

$$Br + H_2 \rightleftharpoons HBr + H \qquad [4.33]$$

is,

$$K_e = k_2/k_5 \qquad (4.55)$$

and may also be calculated from non-kinetic data. Hence,

$$k_5 = k_2 K_e^{-1}$$

and (4.54) gives,

$$= \frac{1}{2}A(K_d)^{-\frac{1}{2}}(K_e)^{-1} \qquad (4.56)$$

From (4.52)

$$k_3 = k_5 B^{-1}$$

$$= \frac{1}{2}(A/B)(K_d)^{-\frac{1}{2}}(K_e)^{-1} \qquad (4.57)$$

Thus experimental measurements on the steady state thermal H_2/Br_2 reaction enable three of the elementary rate constants, i.e. k_2, k_3 and k_5, to be obtained from (4.54), (4.57) and (4.56) respectively.

The remaining rate constants k_1 and k_4 may be obtained by using a different method of activation. Bromine molecules may be dissociated by the absorption of light and so the reaction can be studied photochemically at temperatures where the thermal reaction is negligible. Figure 4.3 shows qualitatively some of the lower potential energy curves for bromine.

Absorption of a quantum of light of suitable energy can induce transition from the ground state to the $^3\Pi_{ou}^+$ state. Collision induced pre-dissociation of this state leads to the production of two ground state $(^2P_{\frac{3}{2}})$ bromine atoms.* Absorption of light of higher frequency in the continuous region of the spectrum leads directly to dissociation of the upper state into one ground state and one excited state $(^2P_{\frac{1}{2}})$ atom. Provided this excited

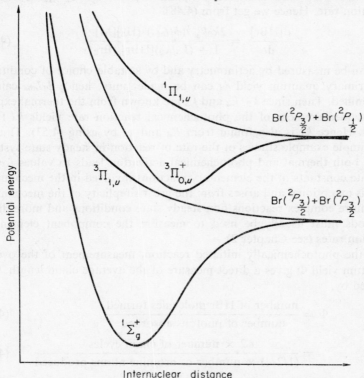

FIGURE 4.3   A schematic potential energy diagram for bromine.

atomic species is rapidly de-activated to the ground state the reaction can be treated as a normal dissociation process. Thus initiation of the photo-chemical chain can be summarized as,

$$Br_2 + h\nu \xrightarrow{6} 2Br \qquad [4.34]$$

with a rate given by,

$$\text{rate} = \phi_i I_{abs} \qquad (4.58)$$

where I_{abs} is the intensity of light absorbed by the bromine (measured in einsteins per second, where one einstein equals one Avogadro's number of

* There is evidence that absorption of wave-lengths between the convergence limit (5,107Å) and 6,800Å leads to the continuum of an attractive $^3\Pi_{1,u}$ state. This forms two $^2P_{\frac{1}{2}}$ bromine atoms directly without molecular collisions. ·

photons) and ϕ_i is the quantum yield for this initiation process, i.e. it is the number of molecules decomposed per photon absorbed. If the photochemical reaction is studied under conditions such that the thermal reaction is negligible the mechanism of the reaction consists of the steps (6), (2), (3), (4) and (5) above. Hence the steady state rate can be derived as before but with the substitution of $2\phi_i I_{abs}$ in place of $2k_1[M][Br_2]$ for the initiation rate. Hence we get from (4.48).

$$\frac{d[HBr]}{dt} = \frac{2k_2 k_4^{-\frac{1}{2}}(\phi_i I_{abs})^{\frac{1}{2}}[H_2][M]^{-\frac{1}{2}}}{1 + (k_5/k_3)[HBr]/[Br_2]} \tag{4.59}$$

I_{abs} can be measured by actinometry and by suitable choice of conditions the primary quantum yield ϕ_i can be made unity, hence $\phi_i I_{abs}$ can be determined. Then since k_2, k_3 and k_5 are known from the thermal experiments, measurement of the photochemical reaction rate yields k_4 from (4.59). Hence k_1 is also found from K_d and k_4 by using (4.53). Thus in this simple example studies of the rate of reaction in steady state systems using both thermal and photochemical activation leads to values for all the rate constants of the elementary reactions involved in the mechanism. This is exceptional and arises from the great simplicity of the mechanism. For more complex reactions non-steady state conditions and more direct methods must usually be used to measure the component elementary reaction rates (see Chapter 2).

In the photochemically initiated reaction, measurement of the overall quantum yield Φ gives a direct measure of the average chain length. Φ is defined by,

$$\Phi = \frac{\text{number of HBr molecules formed}}{\text{number of photons absorbed}} \tag{4.60}$$

$$= \frac{2 \times \text{number of chain cycles}}{(1/2\phi_i) \times \text{number of bromine atoms produced}} \tag{4.61}$$

therefore

$$\Phi = 4\phi_i n_p \tag{4.62}$$

where n_p is the average photochemical chain length, i.e. the average number of chain cycles completed by a single bromine atom. Measurement of this quantum yield shows that for normal light intensities the chain is very short and inefficient. For example at room temperature* Φ is about 10^{-2} under normal experimental conditions of photolysis. This means that most of the bromine atoms produced by the photolysis of bromine molecules simply recombine again without undergoing a propagating reaction at all. Only rarely does a bromine atom manage to complete a

---

* Recent evidence suggests that the bulk of the photochemical reaction at room temperature is a non-chain termolecular reaction $2Br + H_2 \rightarrow 2HBr$, analogous to that of iodine [4.38].

chain cycle. The reason for this inefficiency of the chain becomes clear when the relative magnitudes of the rate constants of the elementary reactions involved are considered. The crucial factor is the relatively high activation energy of the propagating step (2), $E_2 = 17 \cdot 6$ kcal mole^{-1}, which makes this first step of the chain cycle very slow. The second propagating step, reaction (3), is much faster since it has an activation energy of only about 1 kcal mole^{-1}, and so it can be seen from equation (4.40) that $[H] \ll [Br]$. Termination by step (4) is therefore very much faster than by combination of H and Br or of H and H atoms.

The experimental values for the rate constants are approximately,

$$k_1 = 4 \times 10^{15} \exp(-45,000/RT) \text{ cc mole}^{-1} \text{ sec}^{-1}$$
$$k_2 = 4 \cdot 4 \times 10^{13} \exp(-17,600/RT) \text{ cc mole}^{-1} \text{ sec}^{-1}$$
$$k_3 = 1 \cdot 5 \times 10^{14} \exp(-900/RT) \text{ cc mole}^{-1}$$
$$k_4 = 4 \times 10^{15} \text{ cc}^2 \text{ mole}^{-2} \text{ sec}^{-1}$$
$$k_5 = 1 \cdot 8 \times 10^{13} \exp(-900/RT) \text{ cc mole}^{-1} \text{ sec}^{-1}$$

For typical experimental conditions of temperature $= 500°K$ and pressures $= 100$ torr of hydrogen plus 100 torr of bromine, we may calculate when $[HBr] = 0$,

$$[Br] = 1 \cdot 4 \times 10^{-12} \text{ mole cc}^{-1} \text{ or } 4 \cdot 5 \times 10^{-5} \text{ torr}$$
$$[H] = 2 \times 10^{-20} \text{ mole cc}^{-1} \text{ or } 6 \cdot 4 \times 10^{-13} \text{ torr}$$

and the ratio of concentrations of the two active centres is,

$$\frac{[H]}{[Br]} = 1 \cdot 4 \times 10^{-8}$$

The concentration of hydrogen atoms in equilibrium with molecular hydrogen at this temperature and pressure is given by (4.25) as

$$[H] = 1 \cdot 4 \times 10^{-18} \text{ torr}$$

The steady state concentration of this active centre therefore far exceeds this equilibrium value. This is a common situation in chain reactions of this sort. The thermal chain length n_t is given by substituting (4.49) into (4.30),

$$n_t = \frac{k_2(k_1/k_4)^{\frac{1}{2}}[Br_2]^{\frac{1}{2}}[H_2]}{2k_1[Br_2][M]} \tag{4.63}$$

remembering that two molecules of HBr are formed in each chain cycle. Substitution of the numerical values given above for the rate constants and typical conditions yield,

$$n_t = 1 \cdot 5 \exp(+4,900/RT) \tag{4.64}$$

As the temperature is increased the rate of step (2) increases rapidly but that of step (1) does so more rapidly so that the average thermal chain length decreases slowly as seen from (4.64). At 500°K, (4.64) gives $n_t = 200$ which is much greater than n_p under normal conditions of photolysis. The overall rate of reaction is given by substituting the above rate constant values into (4.49) as

$$\left(\frac{d[\text{HBr}]}{dt}\right)_{t=0} = 8\cdot8 \times 10^{13} \exp(-40{,}000/RT)[\text{Br}_2]^{\frac{1}{2}}[\text{H}_2] \text{ mole cc}^{-1} \text{ sec}^{-1}$$

$$(4.65)$$

The overall experimental activation energy of 40 kcal mole^{-1} is only slightly less than the 45 kcal mole^{-1} activation energy of the first step, the dissociation of bromine, owing to the high activation energy of propagation by step (2).

Substitution of the above rate parameters into the relation (4.22) shows that this condition is more than satisfied under all normal conditions owing to the low rate of propagation by step (2). Consequently the extent of reaction occurring during the induction period is negligibly small and the use of the steady state approximation for this reaction as outlined above is fully justified.

It is interesting to compare this simple chain mechanism for the reaction of hydrogen and bromine with the reactions of the other halogens, iodine and chlorine.

4.3d The hydrogen–iodine reaction

At high temperatures this reaction proceeds by a chain mechanism exactly analogous to that described above for bromine. The main difference arises from the relative inactivity of iodine atoms compared to bromine atoms. Thus the first propagating step, which was slow for bromine becomes in the case of iodine,

$$\text{I} + \text{H}_2 \rightarrow \text{HI} + \text{I} \qquad\qquad [4.35]$$

and is very much slower still, having an activation energy of 33·4 kcal mole^{-1}. Because of this high activation energy for propagation the chain reaction rate only becomes measurable at temperatures in excess of 400°C. However below this temperature reaction still occurs, but it does so by a mainly non-chain mechanism. This non-chain reaction has been attributed to a simple bimolecular reaction

$$\text{H}_2 + \text{I}_2 \rightarrow 2\text{HI} \qquad\qquad [4.36]$$

since it is first order with respect to each reactant and the A factor of its rate constant agrees very well with the predictions of both S.C.T. and T.S.T. for elementary bimolecular reactions. Recent photochemical evidence has shown that the reaction involves iodine atoms rather than iodine molecules. The potential energy curves for iodine are qualitatively similar to those for bromine shown in Figure 4.3. Irradiation with light of 5780Å excites the iodine molecules to the $B^3\Pi_{ou}^+$ state (the seventeenth vibrational level). Collision induced predissociation of this state leads to two ground state iodine atoms. The iodine atoms then react with hydrogen to form hydrogen iodide in a reaction whose order with respect to iodine atoms is found experimentally to be the same as that for the recombination to form iodine molecules. The mechanism can therefore be written,

$$I_2 + h\nu \rightarrow 2I \qquad\qquad [4.37]$$

$$2I + H_2 \rightarrow 2HI \qquad\qquad [4.38]$$

$$2I + M \rightarrow I_2 + M \qquad\qquad [4.39]$$

The reaction [4.38] is sufficiently fast to account quantitatively for all of the observed non-chain reaction in the thermally activated systems and so the rate of the bimolecular step [4.36] is negligible in these systems also. In the thermally activated system the establishment of the equilibrium,

$$M + I_2 \rightleftharpoons 2I + M \qquad\qquad [4.40]$$

for which,

$$\frac{[I]^2}{[I_2]} = K \qquad\qquad (4.66)$$

makes any kinetic distinction between [4.36] and [4.38] impossible since from (4.66)

$$k[I_2][H_2] = (k/K)[I]^2[H_2] \qquad\qquad (4.67)$$

In the photochemical experiments however the equilibrium [4.40] is displaced by the absorption of light [4.37] and so the third-order nature of the reaction can be established. Owing to the equilibrium [4.40] the calculation of the rates of [4.36] and [4.38] in thermal systems by T.S.T. yield identical results, (provided the same transition state is assumed for both) as a consequence of the 'equilibrium hypothesis' (section 3.8a). As stated above very good agreement with experiment can be obtained. T.S.T. cannot however be applied to the photochemical system owing to the dis-equilibrium that exists between the reactant states (see p. 145).

Further evidence for the slowness of propagating steps involving iodine atoms such as [4.35] comes from the reverse reaction, the decomposition of hydrogen iodide. This reaction may be brought about photochemically. Figure 4.4 shows qualitatively the potential energy curves for the lower electronic levels of HI. As can be seen from this diagram photolysis of hydrogen iodide at long wavelengths leads to ground state H and I atoms. At somewhat shorter wavelengths it produces excited iodine atoms in the

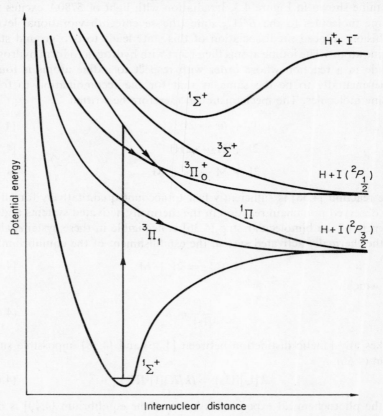

$^1\Sigma^+$

$^3\Sigma^+$

$^3\Pi_0^+$

$^1\Pi$

$^3\Pi_1$

$^1\Sigma^+$

$H^+ + I^-$

$H + I(^2P_{\frac{1}{2}})$

$H + I(^2P_{\frac{3}{2}})$

Potential energy

Internuclear distance

FIGURE 4.4 A schematic potential energy diagram for hydrogen iodide.

$(5^2P_{\frac{1}{2}})$ state and at shorter wavelengths still, ground state $(5^2P_{\frac{3}{2}})$ iodine atoms again. In all cases owing to the conservation of linear momentum most of the excess energy appearing as relative translational motion of the fragments does so as motion of the light hydrogen atom and very little as motion of the very heavy iodine atom. In this way 'hot' hydrogen atoms, (i.e. atoms with translational kinetic energies far in excess of the thermal equilibrium value) with energies of 62 to 84 kcal mole⁻¹ are produced. The overall quantum yield of the decomposition is found to be accurately $2 \cdot 0$ over the whole range of conditions and at all frequencies. The mechanism is,

$$HI + h\nu \rightarrow H + I \qquad [4.41]$$

$$H + HI \rightarrow H_2 + I \qquad [4.42]$$

$$2I + M \rightarrow I_2 + M \qquad [4.43]$$

There is no chain propagated owing to the high activation energy of,

$$I + HI \rightarrow I_2 + H \qquad [4.44]$$

for which $\Delta H = 35$ kcal mole^{-1}, and so the quantum yield does not exceed two. Even the electronically excited $(5^2P_{\frac{1}{2}})$ iodine atoms do not have sufficient energy to react by [4.44] since they are only $21 \cdot 8$ kcal mole^{-1} above the ground state. All the hydrogen atoms formed in [4.41] react by the very fast reaction [4.42] whose activation energy is close to zero and this is therefore true whether or not they are 'hot' species.

4.3e The hydrogen–chlorine reaction

This reaction follows the same basic chain mechanism as the reaction of bromine but it differs from the case of bromine in that the chlorine atom chain centres are much *more* reactive than bromine atoms. Thus the chain propagating step,

$$Cl + H_2 \rightarrow HCl + H \qquad [4.45]$$

which is approximately thermoneutral has an activation energy of only 6 kcal mole^{-1} and is therefore almost as fast as the second propagating step,

$$H + Cl_2 \rightarrow HCl + Cl \qquad [4.46]$$

which has an activation energy of $3 \cdot 0$ kcal mole^{-1}. This makes the chain process very efficient and very long chain lengths are therefore observed. For example, the photochemical reaction at room temperature in the absence of traces of oxygen may have overall quantum yields of up to 10^6. Owing to this great chain length the reaction rate is very sensitive to light which generates chlorine atoms and also to any chemical sensitizers which generate small concentrations of active centres. For the same reason the rate is also very sensitive to traces of inhibitors which react with and destroy the active centres. For example oxygen in very small traces inhibits the reaction owing to the occurrence of the termination processes,

$$H + O_2 + M \rightarrow HO_2' + M \qquad [4.47]$$

$$Cl + O_2 + M \rightarrow ClO_2' + M \qquad [4.48]$$

which generate relatively stable radicals in place of active centres. This makes quantitative study of the reaction extremely difficult since trace amounts of oxygen and other impurities are difficult to remove from the reactants. Wall termination is important and wall initiation possibly also, so that the rate depends markedly on the surface to volume ratio of the reaction vessel.

A further complication in this reaction is that owing to the very fast propagation rate much of the reaction occurs during the induction period before the steady state is reached. This may be seen from the criterion (4.22) by taking the experimental values $k_p = 2 \times 10^{14} \exp(-6,000/RT)$ cc mole^{-1} sec^{-1} and $k_t = 4 \times 10^{15}$ cc^2 mole^{-2} sec^{-1} for [4.45] and the recombination of chlorine atoms respectively, and the typical conditions $T = 500°K$, $P = 100$ torr. Thus substitution into the left hand side of (4.22) gives

$$k_p/4k_t[M] = 6$$

which is two orders of magnitude larger than required by (4.22). However, if sufficient inhibitor is added to the reaction mixture the rate of termination may be increased sufficiently to result in the steady state being reached early in the reaction. For example if oxygen is added in sufficient quantities, reactions [4.47] and [4.48] may become the major termination steps and these it will be noticed are linear terminations. Equation (4.6) then gives the extent of reaction during the induction period as,

$$\ln \frac{[H_2]_0}{[H_2]_{t_i}} = \frac{r_i k_p}{e k_t^2} = \frac{2k_i[M][Cl_2]k_p}{e(k_t'[O_2][M])^2} \qquad (4.68)$$

Taking the approximate values $k_i = 10^{16} \exp(-58,000/RT)$ cc mole^{-1} sec^{-1}, $k_t' = 10^{16}$ cc^2 mole^{-2} sec^{-1} with k_p and the typical conditions as before with an oxygen pressure of 1 torr (1%) we get $\ln[H_2]_0/[H_2]_{t_i} = 10^{-16}$. Hence the inhibited reaction occurs almost entirely in the steady state and the rate expression for the mechanism

$$Cl_2 + M \xrightarrow{\ 1\ } 2Cl + M \qquad\qquad [4.49]$$

$$Cl + H_2 \xrightarrow{\ 2\ } HCl + H \qquad\qquad [4.45]$$

$$H + Cl_2 \xrightarrow{\ 3\ } HCl + Cl \qquad\qquad [4.46]$$

$$H + O_2 + M \xrightarrow{\ 4\ } HO_2^{\cdot} + M \qquad\qquad [4.47]$$

$$Cl + O_2 + M \xrightarrow{\ 5\ } ClO_2^{\cdot} + M \qquad\qquad [4.48]$$

can be found as follows. In the steady state,

$$\frac{d[Cl]}{dt} = 0 = 2r_1 - r_2 - r_5 + r_3 \tag{4.69}$$

$$\frac{d[H]}{dt} = 0 = r_2 - r_3 - r_4 \tag{4.70}$$

Solving these two simultaneous equations for [H] and [Cl] we get,

$$[H] = k_2[Cl][H_2]/(k_3[Cl_2] + k_4[O_2][M])$$

and

$$[Cl] = \frac{2k_1[Cl_2][M]}{k_2[H_2] + k_5[O_2][M] - k_2k_3[H_2][Cl_2]/(k_3[Cl_2] + k_4[O_2][M])} \tag{4.71}$$

The rate of reation is,

$$\frac{1}{2}\frac{d[HCl]}{dt} = \tfrac{1}{2}(r_2 + r_3)$$

$$= \tfrac{1}{2}[Cl]\{k_2[H_2] + k_2k_3[H_2][Cl_2]/(k_3[Cl_2] + k_4[O_2][M])\}$$

Substituting for [Cl] from (4.71) and rearranging gives,

$$\frac{1}{2}\frac{d[HCl]}{dt} = \frac{k_1k_2[H_2][Cl_2][M](1 + 2k_3[Cl_2]/k_4[O_2][M])}{k_2[H_2] + (k_3k_5/k_4)[Cl_2] + k_5[O_2][M]} \tag{4.72}$$

Assuming that the chains are long so that the rates of propagation reactions are much greater than those of termination this expression may be simplified by omitting the term unity in the numerator bracket and the last term in the denominator so that,

$$\frac{1}{2}\frac{d[HCl]}{dt} = \frac{2(k_1k_3/k_4)[H_2][Cl_2]^2}{[O_2]\{[H_2] + (k_3k_5/k_2k_4)[Cl_2]\}} \tag{4.73}$$

This result is in agreement with the experimental rate law for the reaction inhibited by oxygen.

4.3f The pyrolysis of ethane

This reaction has been chosen for discussion since it is an example of a pyrolysis of an organic compound whose mechanism is now well understood as a result of recent analytical experimental work. In the early stages of this pyrolysis the stoichiometry is represented to an accuracy of about 1 % by the simple equation,

$$C_2H_6 = C_2H_4 + H_2 \tag{4.50}$$

As the reaction proceeds however and the product ethylene accumulates the reaction becomes far more complex owing to polymerization and decomposition reactions involving this relatively reactive product and the radicals which propagate the chain decomposition of ethane. For example, reactions such as,

$$C_2H_5^{\cdot} + C_2H_4 \rightarrow C_4H_9^{\cdot} \qquad [4.51]$$

$$C_2H_4 + C_4H_9^{\cdot} \rightarrow C_6H_{13}^{\cdot} \qquad [4.52]$$

$$C_6H_{13}^{\cdot} \rightarrow C_3H_6 + n\text{-}C_3H_7^{\cdot} \qquad [4.53]$$
$$\text{etc.}$$

produce a very complex mixture of products later on in the reaction, and the kinetics of the reaction are correspondingly complex. However, provided the reaction is studied in the very early stages only, before the product ethylene has exceeded about $\frac{1}{2}\%$ of the reactant concentration, the kinetics are found to be relatively simple and can be completely accounted for in terms of the following two centre straight chain scheme,

$$
\left.
\begin{aligned}
C_2H_6 &\xrightarrow{\ 1\ } 2CH_3^{\cdot} \\
CH_3^{\cdot} + C_2H_6 &\xrightarrow{\ 2\ } CH_4 + C_2H_5^{\cdot}
\end{aligned}
\right\} \text{ initiation}
\qquad
\begin{aligned}
&[4.54] \\[4pt]
&[4.55]
\end{aligned}
$$

$$
\left.
\begin{aligned}
C_2H_5^{\cdot} &\xrightarrow{\ 3\ } C_2H_4 + H \\
H + C_2H_6 &\xrightarrow{\ 4\ } C_2H_5^{\cdot} + H_2
\end{aligned}
\right\} \text{ propagation}
\qquad
\begin{aligned}
&[4.56] \\[4pt]
&[4.57]
\end{aligned}
$$

$$
\left.
\begin{aligned}
2C_2H_5^{\cdot} &\xrightarrow{\ 5a\ } n\text{-}C_4H_{10} \\
&\xrightarrow{\ 5b\ } C_2H_4 + C_2H_6 \\
H &\xrightarrow{\ 6\ } \text{wall}
\end{aligned}
\right\} \text{ termination}
\qquad
\begin{aligned}
&[4.58] \\[4pt]
&[4\ 59] \\[4pt]
&[4.60]
\end{aligned}
$$

The C–C bond is the weakest in ethane, with $D(\text{C–C}) \simeq 90$ kcal mole^{-1}, so that production of methyl radicals by step (1) is the most important initiation reaction and its high-pressure rate constant is, $k_1^{\infty} = 10^{16\cdot5}$ $\exp(-88,000/RT)$ sec^{-1}, (see p. 146). The methyl radicals so generated react very rapidly by abstracting a hydrogen atom from an ethane molecule in step (2) to produce ethyl radicals which propagate the chain by step (3). The rate constant for this abstraction (2) is $k_2 = 10^{11} \exp(-10,000/RT)$ cc mole^{-1} sec^{-1} (see p. 92). The ethyl radical can undergo relatively easy unimolecular decomposition to ethylene and a hydrogen atom as in step (3) with a high-pressure rate constant of $k_3^{\infty} = 10^{13} \exp(-40,000/RT)$ sec^{-1} (see pp. 113, 146). The alternative reaction for ethyl, that is hydrogen abstraction from ethane,

$$C_2H_5^{\cdot} + C_2H_6 \rightarrow C_2H_6 + C_2H_5^{\cdot} \qquad [4.61]$$

is in this case chemically undetectable. The chain cycle is completed by the second active centre, H, abstracting a hydrogen atom from the reactant in step (4) to regenerate an ethyl radical chain centre with a rate constant $k_4 = 10^{12} \exp(-6,000/RT)$ cc mole^{-1} sec^{-1} (see p. 92).

The least reactive radical present is ethyl and it might be expected that the concentration of this would far exceed those of the more reactive H and $CH_3^.$ in the steady state. This conclusion is confirmed by the steady state calculations below. Since radical recombination rate constants do not vary greatly from species to species the prime factors in determining the relative rates of the various possible mutual terminations are the concentrations of the various radicals. Termination in the gas phase is therefore predominantly by mutual destruction of two ethyl radicals as in step (5). This can lead either to the formation of normal butane by combination [4.58] or to ethylene and ethane by disproportionation (H transfer) [4.59]. The ratio of the rates of these two processes, the disproportionation/combination ratio, Δ is $0 \cdot 14$ independent of temperature, and the total termination rate constant is $k_5 = 10^{13.25}$ cc mole^{-1} sec^{-1}. Termination by diffusion of the very reactive hydrogen atoms to the wall of the reaction vessel as in step (6) is only important at low pressures. For example in vessels of normal dimensions (6) only becomes significant at pressures below about 10 torr. At higher pressures for simplicity we can ignore (6). Use of equation (4.21) with the above rate constants shows that the extent of reaction that occurs during the induction period is negligible, for example about $10^{-3}\%$ for typical conditions of $T = 900°K$, $P = 100$ torr, and so we may apply the steady state method as follows,

$$\frac{d[CH_3^.]}{dt} = 0 = 2r_1 - r_2 \tag{4.74}$$

$$\frac{d[C_2H_5^.]}{dt} = 0 = r_2 - r_3 + r_4 - 2r_5 \tag{4.75}$$

$$\frac{d[H]}{dt} = 0 = r_3 - r_4 \tag{4.76}$$

Adding these equations we get the usual result for straight chains, that initiation and termination are balanced in the steady state,

$$0 = r_1 - r_5 \tag{4.77}$$

therefore

$$[C_2H_5^.] = \left(\frac{k_1[C_2H_6]}{k_5}\right)^{\frac{1}{2}} \tag{4.78}$$

The rate of formation of the major product hydrogen is given by,

$$\frac{d[H_2]}{dt} = r_4$$

which from (4.76) is,

$$\frac{d[H_2]}{dt} = r_3$$

Substitution for $[C_2H_5^{\cdot}]$ from (4.78) gives,

$$\frac{d[H_2]}{dt} = k_3 \left(\frac{k_1}{k_5}\right)^{\frac{1}{2}} [C_2H_6]^{\frac{1}{2}} \tag{4.79}$$

Also from (4.76) we get,

$$\frac{[H]}{[C_2H_5^{\cdot}]} = \frac{k_3}{k_4}[C_2H_6]^{-1} \tag{4.80}$$

Substituting the numerical values given above for the rate constants gives,

$$\frac{[H]}{[C_2H_5^{\cdot}]} = 10 \exp(-34{,}000/RT)[C_2H_6]^{-1}$$

which for the typical conditions of $T = 900°K$ and $P = 100$ torr is,

$$\frac{[H]}{[C_2H_5^{\cdot}]} \simeq 10^{-2}$$

Similarly from (4.74),

$$[CH_3^{\cdot}] = 2k_1/k_2 \tag{4.81}$$

therefore

$$\frac{[CH_3^{\cdot}]}{[C_2H_5^{\cdot}]} = 2(k_1k_5)^{\frac{1}{2}}/k_2[C_2H_6]^{\frac{1}{2}} \tag{4.82}$$

which for the above conditions gives on substitution,

$$[CH_3^{\cdot}]/[C_2H_5^{\cdot}] \simeq 10^{-2}$$

These results confirm the statement made above that ethyl is the radical which is present in highest concentration under normal conditions.

The rate of formation of the minor product methane is,

$$\frac{d[CH_4]}{dt} = 2k_1[C_2H_6] \tag{4.83}$$

and of n-butane is,

$$\frac{d[C_4H_{10}]}{dt} = \frac{1}{1 + \Delta} r_5$$

where $\Delta = 0.14$. Using equation (4.77) we get,

$$\frac{d[C_4H_{10}]}{dt} = \frac{1}{1 + \Delta} k_1[C_2H_6] \qquad (4.84)$$

Measurement of either of the minor products methane or butane therefore gives the initiation rate directly. The chain length, n, is given by,

$$n = \text{rate of propagation/rate of initiation}$$

so that, $\qquad n = \dfrac{d[H_2]/dt}{d[CH_4]/dt}$

Substitution of (4.79) and (4.83) gives,

$$n = \frac{k_3}{2(k_1k_5)^{\frac{1}{2}}} [C_2H_6]^{-\frac{1}{2}} \qquad (4.85)$$

Experimentally it is found that the methane formed in the early stages is about 1% of the yield of hydrogen under typical conditions. This can be verified from the equation (4.85) by substituting the values for the rate constants given above and leads to the conclusion that $n \simeq 100$. This reaction is therefore a chain of moderate length, and the overall rate of reaction is given by the rate of disappearance of ethane or the rates of production of hydrogen or ethylene to good approximations (within a few per cent) in accordance with [4.50]. Thus,

$$-\frac{d[C_2H_6]}{dt} = r_1 + r_2 + r_4$$

which from (4.74) and (4.76) is

$$= 3r_1 + r_3$$
$$= 3k_1[C_2H_6] + k_3(k_1/k_5)^{\frac{1}{2}}[C_2H_6]^{\frac{1}{2}} \qquad (4.86)$$

$$\frac{d[C_2H_4]}{dt} = r_3 + r_{5b}$$

and from (4.77) and (4.79) together with the ratio $\Delta = r_{5b}/r_{5a}$ gives,

$$\frac{d[C_2H_4]}{dt} = \frac{\Delta}{1 + \Delta} k_1[C_2H_6] + k_3 \left(\frac{k_1}{k_5}\right)^{\frac{1}{2}} [C_2H_6]^{\frac{1}{2}} \qquad (4.87)$$

Both these rates (4.86) and (4.87) are approximately equal to (4.79) to within a few per cent.

$$\frac{d[H_2]}{dt} = k_3(k_1/k_5)^{\frac{1}{2}}[C_2H_6]^{\frac{1}{2}} \qquad (4.79)$$

The overall order of reaction predicted from this result would appear at first glance to be 0.5. However it must be remembered that both steps (1) and (3) are unimolecular decompositions and therefore may have orders anywhere between one, at high pressure, and two at low pressure. In fact for temperatures around $900°K$ and pressures near 100 torr, reaction (1) is close to its high-pressure first-order region but reaction (3) is approximately half-way into its fall-off region. The fall-off of this latter rate constant can be satisfactorily described by Kassel's equation with $n_k = 7$, see equation (3.90). Experimental measurements of the order of step (3) have shown that it is about 1.5 under the conditions quoted above, so that the predicted order for the overall reaction is 1.0 as can be seen from (4.79) if $k_3 \propto P^{0.5}$. The experimental value for this overall order under these conditions is indeed close to unity but of course it varies with conditions, tending to increase at low pressures and decrease at higher pressures. This illustrates how quite a simple integral order for a reaction may be produced by quite a complex mechanism over a limited range of suitable conditions. Rice and Herzfeld first proposed chain schemes of the general type given above for ethane (though they differed somewhat in detail) to explain the simple orders often found for pyrolyses of organic compounds.

The overall activation energy can also be predicted from (4.79). The activation energy for step (3) may be estimated by using equation (3.94) of Kassel's theory and the fact that this reaction is about half-way between first and second order. Thus, we may estimate very approximately that the activation energy is about half-way between its high and low pressure values so that,

$$E_3 \simeq 40 - \tfrac{1}{2}[RT(7 - \tfrac{3}{2})] = 35.5 \text{ kcal mole}^{-1}$$

Whence the overall activation energy is from (4.79), $E_3 + \tfrac{1}{2}[E_1 - E_5] = 35.5 + \tfrac{1}{2} \times 88 = 79.5 \text{ kcal mole}^{-1}$. The experimental value in this pressure range is 78.2 ± 3.0 in good agreement.

At pressures in the region of 50 torr and below the unimolecular dissociation (1) also starts to fall-off and so this causes the overall order of reaction to increase more rapidly as the pressure is reduced. This fall-off can best be studied by measuring the methane formed since from equation (4.83) this gives k_1 directly. It is found that the fall-off in rate can be adequately represented by Kassel's formula (3.90) with $n_k \simeq 12$. In the limit at low pressures where both steps (1) and (2) are second order, equation (4.79) would predict that the overall order of reaction should have risen to 2.0. However, the situation is more complicated than this since, as noted earlier, at pressures much below 10 torr wall termination by reaction (6) becomes significant. If (6) were to become the major termination step at low pressures we may carry out a steady state analysis for the mechanism consisting of reactions (1) to (4) and (6), thus,

$$\frac{d[CH_3^{\cdot}]}{dt} = 0 = 2r_1 - r_2 \qquad (4.88)$$

$$\frac{d[C_2H_5^{\cdot}]}{dt} = 0 = r_2 - r_3 + r_4 \qquad (4.89)$$

$$\frac{d[H]}{dt} = 0 = r_3 - r_4 - r_6 \qquad (4.90)$$

Adding these equations we get,

$$0 = 2r_1 - r_6$$

The rate of reaction (6) can be approximately represented by $k_6\overline{[H]}$, where k_6 depends on the dimensions of the reaction vessel and the diffusion coefficient and $\overline{[H]}$ is a spatially-averaged hydrogen atom concentration. Hence from the above equation,

$$\overline{[H]} = \frac{2k_1[C_2H_6]}{k_6} \qquad (4.91)$$

The rate of formation of hydrogen is,

$$\frac{d[H_2]}{dt} = r_4$$

From (4.91) this is,

$$\frac{d[H_2]}{dt} = \frac{2k_1k_4}{k_6} [C_2H_6]^2 \qquad (4.92)$$

This is an apparently second-order rate equation but as explained above k_1 is pressure dependent and so also is k_6 since the diffusion coefficient is inversely proportional to the pressure for an ideal gas. Thus the overall order predicted by (4.92) could be as high as 4·0 in the extreme case, although this has never been observed experimentally. This treatment based on the expression $r_6 = k_6\overline{[H]}$ is an over-simplification, however more exact computer solution of the diffusion equation for hydrogen atoms has given good agreement with experimental data at low pressures and there is little doubt that the basic mechanism presented here is essentially correct over the whole range of pressures accessible to experiment.

4.3g The pyrolyses of other organic compounds

The same general type of chain mechanism that has been described above in some detail for the case of ethane may also be applied to other thermal decompositions such as those of other paraffins, aldehydes and ethers. In most cases the details are not fully known since insufficient reliable kinetic data is available. However, certain generalizations about these Rice–Herzfeld type mechanisms may be summarized as follows.

The radicals involved as chain carriers in these pyrolyses may be divided into two broad classes called β and μ radicals. The β radicals are those, which like H in the case of ethane pyrolysis, react exclusively in bimolecular propagation steps thus,

$$\beta + A \xrightarrow{k^{\beta}} P + \mu \qquad [4.62]$$

where A is the reactant and P is a product. The μ radicals are those larger radicals which, like C_2H_5 in the case of ethane pyrolysis, react in unimolecular propagation steps thus,

$$\mu + pM \xrightarrow{k_{\mu}} O + \beta + pM \qquad [4.63]$$

where M is an inert energy transfer agent and the coefficient p lies between 0 and 1 and O is an unsaturated product molecule. For long reaction chains in the steady state these two propagating steps are approximately equal in rate since initiation and termination rates are small in comparison with them. Thus,

$$k_{\beta}[\beta][A] = k_{\mu}[\mu][M]^{p} \qquad (4.93)$$

Which reaction is the most important termination step of the chain will depend on the relative concentrations of the β and μ radicals, the total pressure, the vessel dimensions and so on. This reaction may be written in a very general form as,

$$b\beta + m\mu + tM \xrightarrow{k_t} \text{end} \qquad [4.64]$$

where m and b are positive integers or zero and $m + b = 1$ or 2 for linear and quadratic termination respectively. t lies between 0 and 1 for quadratic termination depending to what extent a third body is needed and is -1 for a diffusion controlled linear surface termination such as [4.18] or is between $+1$ and $+2$ for linear gas phase terminations such as [4.13], [4.14] and [4.15]. The initiation reaction may be written,

$$A + iM \xrightarrow{k_i} 2 \text{ active centres} + iM \qquad [4.65]$$

where i lies between 0 and 1. This reaction must balance termination in the steady state for straight chains so that,

$$2k_i[\text{A}][\text{M}]^i = (m + b)k_t[\beta]^b[\mu]^m[\text{M}]^t \tag{4.94}$$

The rate of reaction is equal to the rate of propagation for long chains and hence by solving (4.93) and (4·94) for $[\beta]$ and $[\mu]$ and substituting into either side of (4.93) we get,

$$\text{rate} = \left(\frac{2k_i[\text{M}]^i}{(m + b)k_t[\text{M}]^t}\right)^{1/m+b} k_\beta{}^{b/m+b}k_\mu{}^{m/m+b}[\text{M}]^{mp/m+b}[\text{A}]^{(1+b)/m+b}$$

$$\tag{4.95}$$

From this relation for example we can deduce the overall order of reaction,

$$\text{order} = \frac{(1 + b + mp + i - t)}{(m + b)} \tag{4.96}$$

This relation thus summarizes the steady state solutions for mechanisms of this general type with either first or second (or intermediate) order chain initiation and decomposition of μ radicals, and second or third (or intermediate) order mutual termination involving either $\mu + \mu$, $\mu + \beta$ or $\beta + \beta$ radicals or linear termination involving either β or μ whether on the surface or in the gas phase. To take an example, in the case of ethane at high pressures, $p = 0$, $b = 0$, $m = 2$, $t = 0$, $i = 0$ so that (4.95) becomes,

$$\text{rate} = \left(\frac{k_i}{k_t}\right)^{\frac{1}{2}} k_\mu[\text{A}]^{\frac{1}{2}} \tag{4.97}$$

which agrees with the previous result (4.79). At low pressures where surface termination predominates, $m = 0$, $b = 1$ so that,

$$\text{rate} = \frac{2k_i[\text{M}]^i}{k_t[\text{M}]^t} k_\beta[\text{A}]^2 \tag{4.98}$$

in agreement with (4.92).

4.3h The inhibition of organic pyrolyses

One common characteristic of all pyrolyses of organic compounds that proceed by chain mechanisms of the type outlined above is their susceptibility to inhibition, i.e. a slowing down in rate, by substances such as nitric oxide, propylene and other olefines. At one time it was thought that the action of these inhibitors was to completely suppress the reaction chains and to leave a residual reaction that was entirely molecular in character. It is now known that this is not so. Numerous experimental proofs have been found to show that the inhibited reactions are still chain processes but are more complex than the uninhibited reactions. For example, pyrolysis of deuterated paraffins leads to isotopic mixing in the products occurring via free radical intermediates and the extent of the mixing is unaltered by the addition of nitric oxide.

The inhibited paraffin pyrolyses have been studied to a greater extent than other inhibited reactions. Briefly, the experimental characteristics

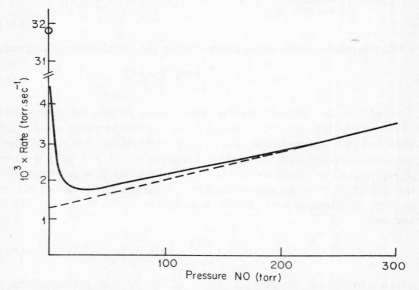

FIGURE 4.5   A typical 'inhibition curve' for the inhibition of the initial rate of formation of hydrogen from ethane (100 torr) by nitric oxide at 597°C. The ordinate is broken as indicated and the uninhibited rate is shown by the circle.

of these inhibitions are as follows. Addition of small amounts of inhibitor reduces the initial rate of decomposition of the paraffin. As the reaction proceeds however it accelerates to a maximum rate before slowing down towards completion. This behaviour may be contrasted to the continuously decelerating rate that is observed for the uninhibited pyrolyses (the induction period for the establishment of the steady state in paraffin pyrolyses is too short to be observable, (see p. 187). Little is known of the reasons for this self-acceleration of the inhibited reaction and the extent of inhibition is usually judged by considering the initial rate only. As increasing amounts of inhibitor are added the initial rate varies in the manner illustrated in Figure 4.5. The initial rate at first decreases rapidly then passes through a flat minimum and then increases at large amounts of inhibitor eventually varying linearly with the 'inhibitor' concentration. The minimum rate is often called the residual or limiting rate. The region of increasing rate is called the induced reaction and was in fact discovered before inhibition, though very little is known about its nature. For many reactions the minimum rate is approximately the same under given conditions irrespective of whether nitric oxide or olefines are used as inhibitors. It was this fact which lead to the earlier hypothesis that fully inhibited reactions are molecular. However this equality of the rates with different inhibitors is only very approximate, typically within about 20%. It is also destroyed by alterations in conditions such as reaction temperature, reaction vessel surface, and reactant pressure. An additional complication is that the olefines themselves pyrolyse rapidly under the conditions in which they are used as inhibitors, and the amounts of these olefinic inhibitors used to reduce the rate to a minimum are comparable to the amounts of the paraffin taken so that the reaction in reality is the co-pyrolysis of two compounds of comparable reactivity. As was noted in the case of the ethylene formed in the ethane pyrolysis (p. 186) the reaction is exceedingly complex even when only a few per cent of olefine is present in the paraffin.

For this reason the mechanisms of the inhibitions produced by nitric oxide have been investigated to a greater extent than those of olefinic inhibitors. As yet no completely satisfactory mechanism has been found even in the simplest cases. A variety of possibilities must be taken into account in devising any complete mechanism of the action of nitric oxide and a few of these will be outlined very briefly here. Firstly it must be recognized that nitric oxide is itself a free radical albeit a relatively stable one. Reactions such as,

$$R\text{--}H + NO \rightleftharpoons R^\cdot + HNO \qquad [4.66]$$

in which the inhibitor *initiates* reaction chains must be considered possible. More important still the reactions of the products formed when nitric oxide combines with active centres in chain termination must be considered. These products are nitroso compounds,

$$\beta + NO + M \rightarrow \beta NO + M \qquad [4.67]$$

$$\mu + NO \rightarrow \mu NO \qquad [4.68]$$

Nitroso compounds isomerize rapidly in certain conditions to form the iso-nitroso compounds or oximes as they are more usually called, for example,

$$C_2H_5NO \rightarrow CH_3CH{=}NOH \qquad [4.69]$$

These oximes themselves decompose very rapidly via free radical chain mechanisms under the conditions of the inhibited pyrolyses since the dissociation energy of the N–O bond, e.g. in,

$$CH_3CH{=}NOH \rightarrow CH_3CHN^{\cdot} + OH^{\cdot} \qquad [4.70]$$

is only about 47 kcal mole^{-1} which is very much less than that of any bond in the paraffins (the dissociation of which initiates the uninhibited chain reactions). Furthermore these oximes undergo very rapid reactions with nitric oxide probably by such processes as for example,

$$CH_3CH{=}NOH + NO \rightarrow CH_3CHN^{\cdot} + HONO \qquad [4.71]$$

$$HONO \rightarrow NO + OH^{\cdot} \qquad [4.72]$$

Processes of this sort can clearly lead to a form of degenerate branching of the reaction chain, (see section 4.4d). Other reactions of nitroso compounds which must be considered are the following steps which are known with some certainty to occur at temperatures near room-temperature,

$$RNO + NO \rightleftarrows R(NO)_2 \qquad [4.73]$$

$$R(NO)_2 + NO \rightarrow R(NO)_3 \qquad [4.74]$$

$$R(NO)_3 \rightarrow RN_2NO_3 \qquad [4.75]$$

$$RN_2NO_3 \rightarrow R^{\cdot} + N_2 + NO_3 \qquad [4.76]$$

$$NO_3 + NO \rightarrow 2NO_2 \qquad [4.77]$$

At the higher temperatures used for paraffin pyrolyses reaction [4.77] must be replaced by,

$$M + NO_3 \rightarrow NO + O_2 + M \qquad [4.78]$$

The production of oxygen in this way would result in the inhibited reactions assuming many of the complexities of combustions (see sections 4.4c and 4.4d).

The corresponding reactions to the above steps in the case of olefinic inhibitors say propylene for example would be,

$$R^{\cdot} + C_3H_6 \rightleftarrows RH + C_3H_5^{\cdot} \qquad [4.79]$$

$$R^{\cdot} + C_3H_6 \rightarrow RC_3H_6^{\cdot} \qquad [4.80]$$

$$C_3H_5^{\cdot} + C_3H_6 \rightarrow C_6H_{11}^{\cdot} \qquad [4.81]$$

$$C_6H_{11}^{\cdot} \rightarrow C_6H_{10} + H \qquad [4.82]$$

$$\text{or } C_5H_8 + CH_3^{\cdot}$$

etc.

together with termination reactions involving the allyl radical $C_3H_5^{\cdot}$. All that can really be said with any certainty about the mechanisms of inhibited pyrolyses is that they are very complex chain processes in which the inhibitor and its reaction products are intimately involved. The use of inhibitors as a diagnostic test for the presence of free radical chain reactions remains a useful technique but attempts to simplify otherwise complex processes by their use to destroy active centres are bound to be unsuccessful. Furthermore lack of inhibition cannot be taken as conclusive proof that free radical chains are not present in a reaction.

4.3i Addition polymerization

As a final example of straight chain mechanisms we will consider the free radical addition polymerization of vinyl compounds of the general formula $R_1R_2C{=}CR_3R_4$. These reactions proceed by the successive addition of monomer molecules to a growing polymer free radical and are very clear examples of chain kinetics since the reactions actually result in the formation of long chain molecules. The number of different chain centres in this type of reaction is virtually infinite since the growing radical may have any length and still be sufficiently reactive to add on another monomer molecule, M, thus,

$$\text{initiator} \xrightarrow{\;r_i\;} R_0^{\cdot} \qquad\qquad \text{initiation} \qquad\qquad [4.83]$$

$$\left.\begin{array}{l} R_0^{\cdot} + M \rightarrow R_1^{\cdot} \\[4pt] R_1^{\cdot} + M \rightarrow R_2^{\cdot} \\[2pt] \quad\vdots \\[2pt] R_n^{\cdot} + M \rightarrow R_{n+1}^{\cdot} \\[2pt] \quad\vdots \end{array}\right\} \text{propagation} \quad \begin{array}{r} [4.84] \\[8pt] [4.85] \\[14pt] [4.86] \end{array}$$

$$\left.\begin{array}{l} R_n^{\cdot} + M \rightarrow R_0^{\cdot} + P_{n+1} \\[4pt] R_n^{\cdot} + M \rightarrow R_1^{\cdot} + P_n \\[2pt] \quad\vdots \\[2pt] \quad\text{etc.} \end{array}\right\} \text{chain transfer} \quad \begin{array}{r} [4.87] \\[8pt] [4.88] \end{array}$$

$$R_n^{\cdot} + R_m^{\cdot} \rightarrow P_{n+m} \qquad\qquad \text{termination} \qquad\qquad [4.89]$$

where $R^{\cdot}$ denotes an active centre and P a product polymer molecule, the suffixes indicating the number of monomer units M in each. The propagation steps such as [4.87], [4.88], etc., that re-generate smaller growing centres are usually called chain transfer steps though it should be noticed that without these steps the mechanism is strictly speaking not a true chain process but simply an infinite sequence of *different* consecutive reactions since at no stage would the initial active centre be regenerated to complete a cycle. The kinetics of these complex chain reactions can clearly be exceedingly complicated. As an example we will consider only a very simple case in which chain transfer is ignored (thus eliminating any chain cycles) and in which the growing active centres are all of very similar reactivity, as may occur for instance if they are all very large. Thus we will assume that the above propagating steps all have the same rate constant k_p, and similarly all the terminating steps have the same rate constant k_t. We will also assume that most of the reaction occurs after the steady state for *all* the active centres has been reached. Hence we may write,

$$\frac{d[R_0^\cdot]}{dt} = 0 = r_i - k_p[R_0^\cdot][M] - k_t[R_0^\cdot]\sum_{n=0}^{\infty}[R_n^\cdot] \qquad (4.99)$$

$$\frac{d[R_1^\cdot]}{dt} = 0 = k_p[R_0^\cdot][M] - k_p[R_1^\cdot][M] - k_t[R_1^\cdot]\sum_{n=0}^{\infty}[R_n^\cdot] \qquad (4.100)$$

$$\frac{d[R_n^\cdot]}{dt} = 0 = k_p[R_{n-1}^\cdot][M] - k_p[R_n^\cdot][M] - k_t[R_n^\cdot]\sum_{n=0}^{\infty}[R_n^\cdot] \qquad (4.101)$$
$$\vdots$$
etc.

Adding all these equations we get the usual result that initiation equals termination thus,

$$0 = r_i - k_t(\sum_{n=0}^{\infty}[R_n^\cdot])^2 \qquad (4.102)$$

since all the propagation steps cancel. The total active centre concentration is, therefore

$$\sum_{n=0}^{\infty}[R_n^\cdot] = (r_i/k_t)^{\frac{1}{2}} \qquad (4.103)$$

The total rate of removal of monomer is,

$$-\frac{d[M]}{dt} = k_p[M]\sum_{n=0}^{\infty}[R_n^\cdot]$$

which from (4.103) is,

$$-\frac{d[M]}{dt} = k_p \left(\frac{r_i}{k_t}\right)^{\frac{1}{2}} [M] \qquad (4.104)$$

The average number of propagating steps performed by a single growing centre is given by $-(d[M]/dt)/r_i$. This is called the 'kinetic chain length' though this name is only consistent with our previous definition of chain length if all the active centres are regarded as being identical, so that the mechanism becomes a one-centre chain and the number of propagating steps completed is the same as the number of complete chain cycles completed. Actually this quantity is the average number of different active centres involved. From (4.104),

$$-\frac{d[M]}{dt}\bigg/ r_i = \frac{k_p}{(r_i k_t)^{\frac{1}{2}}}\,[M] \tag{4.105}$$

Since termination is by the combination* of two active centres [4.89] this average number of propagating steps will also equal one half the average number of monomer units, $\bar{n}$, in the polymer that is produced,

$$-\frac{d[M]}{dt}\bigg/ r_i = \tfrac{1}{2}\bar{n} \tag{4.106}$$

This number may be determined by measuring the molecular weight distribution of the polymer product. All the quantities in equation (4.106) can be directly measured since the rate of initiation can be determined for any particular method of initiation. Thus, for thermal initiation $r_i = 2k_i[M]$ or possibly $r_i = 2k_i[M]^2$; for sensitized initiation by a 'catalyst'[†] C, $r_i = 2k_i[C]$ or possibly $r_i = 2k_i[C][M]$; and for photochemical initiation $r_i = 2\phi_i I_{absorbed}$. These rate parameters may be determined by studies of the thermal decomposition or photolysis of the monomer or sensitizer alone. Chain transfer reactions clearly will have the effect of reducing the molecular weight of the product since they regenerate smaller active centres. Deviations from (4.106) thus give a measure of the extent to which chain transfer processes occur. Studies under non-steady conditions, such as intermittent photolysis (see section 2.5f) can lead to the determination of k_t from the average life time τ of a growing centre.

$$\tau = \sum_{n=0}^{\infty} [R_n^{\cdot}]/r_i$$
$$= (r_i k_t)^{-\frac{1}{2}} \tag{4.107}$$

The propagation rate constant k_p can then be calculated from the observed rate of reaction using equation (4.104).

Having considered some of the properties of straight chain reactions we will now turn to the more complex processes in which branching of the chain occurs.

* Ignoring disproportionation, for simplicity. For reactions in which disproportionation is the major termination the factor of $\frac{1}{2}$ should be left out of (4.106).

† Not a true catalyst since fragments of the sensitizer are incorporated in the polymer produced.

4.4 BRANCHED CHAIN REACTIONS

On page 162 we saw that the propagating steps of a straight chain reaction are always of the general type,

$$R^{\cdot} + \cdots \rightarrow R^{\cdot\prime} + \cdots \qquad [4.90]$$

in which one active centre always generates one, and only one, new active centre. Thus straight propagating steps do not in anyway alter the total concentration of active centres present in the reacting system. However, some reactions of free radicals result in an increase in the number of free radicals in the system thus,

$$R^{\cdot} + \cdots \rightarrow \alpha R^{\cdot\prime} + \cdots \qquad [4.91]$$

where $\alpha > 1$, and is called the branching coefficient. This type of process is called a branching reaction and gives rise to *branching chain* reactions which have some very different properties to those of straight chains. In a qualitative way we can see that if the rate of branching is small then chain termination processes may be able to control the concentration of active centres at a low level and a steady state may be set up in a similar fashion to that which occurs in straight chain reactions. However if the rate of branching is fast it is clear that it may produce radicals at a rate which exceeds that at which they are destroyed by termination. Under these conditions the chain centres may multiply without limit so that the rate of reaction, which depends on the concentration of active centres, also increases without limit and no steady state is reached no matter for how long the reaction proceeds. Such continuously accelerating chemical reactions give rise to the experimental phenomena called explosions. Explosions are of two main types called *thermal explosions* and *isothermal* or *branched chain explosions*. The latter are specific to branched chain reactions while the former may occur in a wide variety of reaction types.

4.4a Thermal explosions

A thermal explosion is one in which the prime cause of the self-acceleration of the reaction is the self-heating of the reaction mixture caused by an exothermic reaction. In general for an exothermic reaction the temperature of an element of the reaction mixture will stabilize when the rate of heat generation by the reaction is balanced by the rate of heat loss from the element by conduction, convection and radiation. The rate of heat production Q_R in a volume of gas V reacting at a rate per unit volume r with an exothermicity $-\Delta H$ per mole of reaction is given by,

$$Q_R = -Vr\Delta H$$

Now $r = kC^x$ where C is the concentration, x is the overall order of reaction and k is the rate constant which depends on temperature according to the Arrhenius law (1.50), $k = A\, e^{-E_a/RT}$. Hence,

$$Q_R = -VC^x A\, e^{-E_a/RT}\Delta H \tag{4.108}$$

The rate of heat loss from the gas depends in a complex way on the physical parameters of the system but for small differences between the gas temperature T and the temperature of the surroundings T_s the rate of heat loss Q_C may be approximately described by the empirical law known as Newton's law of cooling,

$$Q_C = Sh(T - T_s) \tag{4.109}$$

where S is the surface area of the gas element and h is a heat transfer coefficient per unit surface area. The dependence of Q_R and Q_C upon the temperature of the gas is plotted out in Figure 4.6 for three values of the

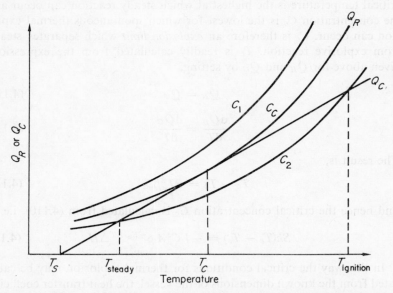

FIGURE 4.6 Rates of heat production Q_R and heat loss Q_C versus temperature for an exothermic reaction at three reactant concentrations $C_1 > C_c > C_2$.

concentration $C_1 > C_c > C_2$. For the lowest of these concentrations C_2 it is clear from the figure that if a gas mixture is made up initially at a temperature T_s, the temperature of the reaction vessel, then Q_R will exceed Q_C and there will be a net heating of the gas mixture and the temperature of the gas will rise until T_{steady} is reached where the two curves of Q_R and

Q_C intersect. At this temperature the system is stable for small displacements since if, for example, the temperature of the gas were to rise above T_{steady} due to some small disturbance then Q_C would exceed Q_R and the gas would tend to cool down to T_{steady} again. Similarly, if a small temperature drop was caused the system would tend to return to T_{steady} since Q_R would exceed Q_C. If a large upwards displacement of gas temperature were to take place to $T_{ignition}$ then the system would cease to be stable. For above $T_{ignition}$, the ignition temperature, Q_R exceeds Q_C again and so the temperature of the gas will increase without limit, or rather in practice until consumption of the reactant reduces the reaction rate sufficiently. At the highest concentration plotted in Figure 4.6, C_1, the curve for Q_R lies wholly above that for Q_C. Hence ignition will occur when the gas mixture is introduced to the vessel at T_s. The intermediate limiting case for concentration C_c occurs when the two curves touch at the critical temperature, T_c where T_{steady} and $T_{ignition}$ coincide. This critical temperature is the highest at which steady reaction can occur and the concentration C_c is the lowest for which spontaneous thermal explosion can occur. C_c is therefore an *explosion limit* which separates steady from explosive reaction. T_c is readily calculated from the expressions given above for Q_R and Q_C by setting,

$$Q_R = Q_C \tag{4.110}$$

$$\frac{dQ_R}{dT} = \frac{dQ_C}{dT} \tag{4.111}$$

The result is,

$$T_c - T_s = RT_c^2/E_a \tag{4.112}$$

and hence the critical concentration C_c is calculated from (4.110), i.e.

$$Sh(T_c - T_s) = -VC_c^x A\, e^{-E_a/RT}\Delta H \tag{4.113}$$

In this way the critical conditions for thermal explosion may be calculated from the known dimensions of the vessel, the heat transfer coefficient and the kinetic and thermodynamic parameters for the reaction. Clearly any exothermic reaction with a positive activation energy can undergo thermal explosion under suitable conditions. For example the explosions that occur in the hydrogen–chlorine reaction are thermal in nature. More rigorous treatments than that used above which take into account the temperature gradients which exist in the reacting gas give more accurate expressions for the explosion limit but do not alter the essence of the general conclusions reached here. The principal reason for outlining the theory of thermal explosions here is to distinguish these from the isothermal explosions of branched chain reactions.

4.4b Isothermal explosions

An isothermal explosion is one in which the prime cause of the self-acceleration is the multiplication of active centres due to a branching chain reaction. The term is not meant to imply, of course, that the actual explosion process is really isothermal but simply that the system would still explode even if it were possible by some means to keep the temperature of the reactants constant. This is because the fundamental reason for the occurrence of the explosion is chemical not thermal in nature. When the branched chain process becomes self-accelerating in this way then, for an exothermic reaction, there obviously is a point at which self-heating becomes important and the final explosion which occurs involves both thermal and chemical effects. Nevertheless the prime cause remains the chemical one and so the characteristics of isothermal explosions, the explosion limits, etc., are very different from those of thermal explosions that were discussed in section 4.4a. To account for isothermal explosions we will apply the quasi-stationary state method outlined previously (sections 1.6c and 4.3a) to a general type of branched chain mechanism. Because of the quasi-stationary state assumption we need only consider the total active centre concentration $[R^{\cdot}]$ and therefore, as before, we will write the mechanism without distinguishing one type of radical from another as follows,

$$M_1 \xrightarrow{\ r_i\ } R^{\cdot} \qquad \text{initiation} \qquad [4.92]$$

$$R^{\cdot} + M_2 \xrightarrow{\ r_p\ } M_3 + R^{\cdot} \qquad \text{propagation} \qquad [4.93]$$

$$R^{\cdot} + M_4 \xrightarrow{\ r_b\ } \alpha R^{\cdot} \qquad \text{linear branching} \qquad [4.94]$$

$$R^{\cdot} + R^{\cdot} \xrightarrow{\ r'_b\ } \alpha' R^{\cdot} \qquad \text{quadratic branching} \qquad [4.95]$$

$$R^{\cdot} + M_5 \xrightarrow{\ r_{tg}\ } R_{\text{inactive}} \qquad \text{linear gas-phase termination} \qquad [4.96]$$

$$R^{\cdot} + R^{\cdot} + M \xrightarrow{\ r'_{tg}\ } M_6 + M \qquad \text{quadratic gas-phase termination} \qquad [4.97]$$

$$R^{\cdot} \xrightarrow{\ r_{ts}\ } \text{wall} \qquad \text{linear surface termination} \qquad [4.98]$$

where $R^{\cdot}$ is an active centre, M_i is any molecule and α and α' are the linear and quadratic branching coefficients respectively. The rate of change of the total active centre concentration in the quasi-stationary state is,

$$\frac{d[R^{\cdot}]}{dt} = r_i + (\alpha - 1)r_b - r_{tg} - r_{ts} + (\alpha' - 2)r'_b - r'_{tg} \qquad (4.114)$$

The second, third and fourth terms of the right hand side of this equation are linear terms ($\propto [R^{\cdot}]$) while the fifth and sixth are quadratic ($\propto [R^{\cdot}]^2$), hence we may re-write this equation as,

$$\frac{d[R^{\cdot}]}{dt} = r_i + \phi[R^{\cdot}] + \phi'[R^{\cdot}]^2 \qquad (4.115)$$

where,

$$\phi = [(\alpha - 1)r_b - r_{tg} - r_{ts}]/[R] \qquad (4.116)$$

is called the *net linear branching factor*, and,

$$\phi' = [(\alpha' - 2)r_b' - r_{tg}']/[R^{\cdot}]^2 \qquad (4.117)$$

is called the *net quadratic branching factor*.

For the sake of simplicity we will consider the case of a linearly branched, linearly terminated chain only. For this type of reaction $\phi' = 0$ and equation (4.115) simplifies to,

$$\frac{d[R^{\cdot}]}{dt} = r_i + \phi[R^{\cdot}] \qquad (4.118)$$

which with the initial condition that $[R^{\cdot}] = 0$ at $t = 0$, integrates to,

$$[R^{\cdot}] = \frac{r_i}{\phi}(e^{\phi t} - 1) \qquad (4.119)$$

as is easily verified by differentiation. For long reaction chains the rate of reaction is equal to the rate of propagation hence,

$$\text{rate} = k_p[R^{\cdot}][M_2]$$

$$= k_p[M_2]\frac{r_i}{\phi}(e^{\phi t} - 1) \qquad (4.120)$$

The variations of the rate and $[R^{\cdot}]$ with time are seen from (4.119) and (4.120) to fall into two clear-cut cases depending on whether ϕ is greater or less than zero. These are illustrated in Figure 4.7. When $\phi < 0$ (4.120) becomes,

$$\text{rate} = \frac{r_i k_p[M_2]}{|\phi|}(1 - e^{-|\phi|t}) \qquad (4.121)$$

where $|\phi|$ is the magnitude of ϕ, i.e. a positive number. This shows the normal behaviour of a reaction approaching a steady state [compare equation (4.3)] as discussed for straight chain reactions in section 4.3a. The induction period is given by $|\phi|^{-1}$. When $\phi > 0$, (4.120) shows that the variation of rate with time follows a continuously accelerating, exponential, rise as shown in Figure 4.7. The sharp division between these two types of behaviour is expressed by the condition,

$$\phi = 0 \qquad\qquad (4.122)$$

This shows that there is an explosion limit at which a very small change in conditions, such as the temperature, pressure or composition of the reactant, can change the reaction rate from a slow steady value to one which is self-accelerating and would eventually become infinitely fast, other factors

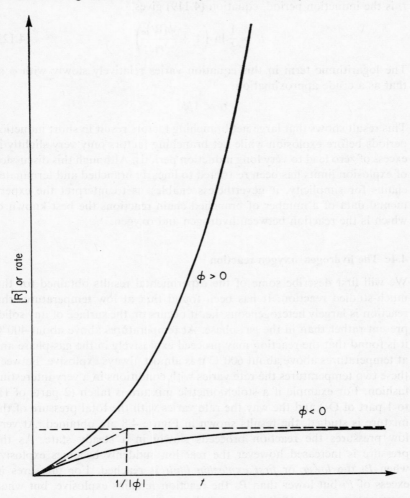

FIGURE 4.7 A plot of $(e^{\phi t} - 1)/\phi$ versus t for $\phi > 0$ (+1) and $\phi < 0$ (−1) showing the difference between steady and explosive reaction.

remaining unchanged. When the rate of reaction reaches a certain value
the observable characteristics of an explosion appear such as the shock
wave causing a bang to be heard, the high temperature causing a flash of
light to be seen, etc. There is however a time-lag before the rate reaches this
critical value, that is there is an induction period before explosion. This
may be of the order of milliseconds or minutes depending on the value of
ϕ. Let $[R^{\cdot}]_c$ be the critical active centre concentration at which the rate is
just fast enough to produce the observable criteria of explosion. Then if
t_i is the induction period, equation (4.119) gives,

$$t_i = \frac{1}{\phi} \ln \left(1 + \frac{\phi [R^{\cdot}]_c}{r_i} \right) \tag{4.123}$$

The logarithmic term in this equation varies relatively slowly with ϕ so
that as a crude approximation,

$$t_i \propto 1/\phi$$

This result shows that large net branching factors result in short induction
periods before explosion while net branching factors only very slightly in
excess of zero lead to very long induction periods. Although this discussion
of explosion limits has been restricted to linearly branched and terminated
chains for simplicity, it nevertheless enables us to interpret the experimental data of a number of branched chain reactions the best known of
which is the reaction between hydrogen and oxygen.

4.4c The hydrogen–oxygen reaction

We will first describe some of the experimental results obtained for this
much-studied reaction. It has been found that at low temperatures this
reaction is largely heterogeneous, i.e. it occurs on the surface of any solids
present rather than in the gas-phase. At temperatures above about 400°C
it is found that the reaction may proceed explosively in the gas-phase and
at temperatures above about 600°C it is almost always explosive. Between
these two temperatures the rate varies with conditions in a very interesting
fashion. For example if a stoichiometric mixture is taken (2 parts of H_2
to 1 part of O_2) and the way the rate varies with the total pressure of the
mixture is studied, the results shown in Figure 4.8 are obtained. At very
low pressures the reaction proceeds slowly in a steady state. As the
pressure is increased however the reaction suddenly becomes explosive
when P_l the *lower* or *first explosion limit* is reached. For pressures in
excess of P_l but lower than P_u the reaction remains explosive, but when
pressures in excess of P_u are used the reaction suddenly becomes non-
explosive again and proceeds in a steady state. P_u is called the *upper* or

second explosion limit. Further increase of pressure eventually leads to the reaction again becoming explosive at much higher pressures at the *third limit* P_t. This behaviour is contrasted in Figure 4.8 with the behaviour to be expected for a 'normal' exothermic reaction undergoing thermal explosion at a single thermal limit. The thermal limit to be expected for this reaction lies somewhat above the observed third limit. These three

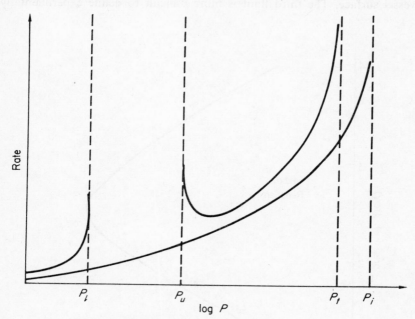

FIGURE 4.8   The variation of the rate of the hydrogen–oxygen reaction with pressure for a stoichiometric mixture (upper line) compared with that to be expected for a non-branching exothermic reaction undergoing thermal explosion at the limit P_i (lower line). P_l, P_u and P_t are the lower, upper and third limits respectively.

limits may be plotted as a function of temperature and pressure on what is called an explosion diagram. The explosion diagram for stoichiometric hydrogen–oxygen mixtures in a KCl-coated vessel of 7·4 cm diameter is shown in Figure 4.9. The area to the left of the line represents the non-explosive reaction region while the area to the right of the line is the explosive region. The area between the lower and upper limits is referred to as the explosion peninsula. The lower limit is seen to decrease slowly as the temperature is raised. It is also lowered by the addition of inert gases to the reaction mixture. It is very sensitive to the nature of the surface of the reaction vessel and is inversely proportional to the diameter of the vessel. The upper limit increases rapidly with increasing temperature and a plot of $\ln P_u$ versus $(1/RT)$ yields an apparent activation energy of between 20 and 25 kcal mole^{-1}. The second limit also is decreased by the addition of inert gases and water is found to be especially effective in this respect. It is almost unaffected by the nature and extent of the reaction vessel surface. The third limit is more difficult to define experimentally

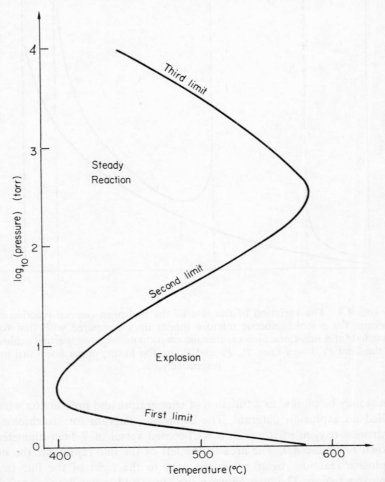

FIGURE 4.9   The explosion limit diagram for a stoichiometric mixture of hydrogen and oxygen in a spherical KCl-coated vessel of 7·4 cm diameter (Lewis and Von Elbe, *Combustion, Flames, and Explosions in Gases*, Academic Press, New York, 1951).

since the reaction rates are very rapid in the non-explosive region and the limit is close to the thermal limit so that considerable self-heating occurs. The third limit decreases rapidly with increasing temperature and is lowered by adding inert gases and by increasing the diameter of the vessel. These effects are summarized in Table 4.4.

TABLE 4.4 The effect of temperature, surface to volume ratio, and inert gas pressure on the explosion limits of the hydrogen–oxygen reaction.

| Limit | Increase of | | |
	Temperature	Surface to volume ratio	Inert gas pressure
First (lower)	lowers	raises	lowers
Second (upper)	raises	has no effect	lowers
Third	lowers	raises	lowers

All the features described above can be interpreted in terms of a branched chain mechanism. Although some details of the mechanism remain obscure the main features are clear. Spectroscopic, mass-spectroscopic and other direct observational techniques (see Chapter 2) have shown the presence in the reaction of the species, H, OH·, O, and HO_2. The exact nature of the chain initiation reaction is uncertain. For simplicity we will write it as,

$$H_2 + S \xrightarrow{r_i} 2H + S \qquad \Delta E = +102 \qquad [4.99]$$

where S represents the wall of the reaction vessel. The branching steps in the mechanism are known to be,

$$H + O_2 \xrightarrow{r_b} OH· + O \qquad \Delta E = +18 \qquad [4.100]$$

$$O + H_2 \longrightarrow OH· + H \qquad \Delta E = +4 \qquad [4.101]$$

This is linear branching with a branching coefficient $\alpha = 3$, since one hydrogen atom as a result of these two reactions produces two OH· radicals and another hydrogen atom. Straight chain propagation occurs by,

$$OH· + H_2 \xrightarrow{r_p} H_2O + H \qquad \Delta E = -17 \qquad [4.102]$$

The main gas-phase termination reaction is,

$$H + O_2 + M \xrightarrow{k_{tg}} HO_2^{\cdot} + M \qquad [4.103]$$

which is followed by,

$$HO_2^{\cdot} \rightarrow \text{wall} \qquad [4.104]$$

The relatively stable radical $HO_2^{\cdot}$ has a life-time of about 10^{-5} sec under typical conditions. Surface termination is also important and may be written,

$$H \text{ or } OH^{\cdot} \xrightarrow{k_{ts}} \text{wall} \qquad [4.105]$$

Both termination steps are linear and we may therefore apply our simple model of section 4.4b for a linearly branched, linearly terminated chain in the quasi-stationary state in which, $[OH^{\cdot}] \propto [H] \propto [O]$. The net branching factor is given by (4.116) and is this case substitution of the rate expressions gives,

$$\phi = 2k_b[O_2] - k_{tg}[O_2][M] - k_{ts} \qquad (4.124)$$

Considering the case of a reaction mixture of fixed stoichiometry, we may write,

$$\frac{[O_2]}{o} = \frac{[H_2]}{h} = \frac{[M]}{m} = P \qquad (4.125)$$

where P is the total pressure and o, h, and m are constants. Now k_{ts} is the rate constant for diffusion of active centres to the wall of the reaction vessel. The rate of this process is proportional to the diffusion coefficient and inversely proportional to the square of the vessel diameter d (compare Einstein's diffusion equation $x^2 = 2Dt$). The diffusion coefficient is equal to the diffusion coefficient at unit pressure D_0 divided by the total pressure. Thus we may put,

$$k_{ts} = \frac{kD_0}{d^2P} \qquad (4.126)$$

Equation (4.124) becomes,

$$\phi = 2k_boP - k_{tg}omP^2 - \frac{kD_0}{d^2P} \qquad (4.127)$$

The explosion limit condition is (4.122) $\phi = 0$, hence at the limit P_c,

$$0 = 2k_boP_c - k_{tg}omP_c^2 - \frac{kD_0}{d^2P_c} \qquad (4.128)$$

This is a cubic equation for P_c and as can be seen from the signs of the terms, two of the roots are positive and one negative. The latter cannot correspond to any physically observable system but the two positive roots account for the lower and upper explosion limits that are observed. At the lower limit P_l, the principal termination reaction is the surface one and the second term of the right-hand-side of (4.128), the gas-phase termination, is small so that we have approximately,

$$0 \simeq 2k_{b}oP_l - \frac{kD_0}{d^2P_l}$$

or,

$$P_l \simeq \frac{1}{d}\left(\frac{kD_0}{2k_{b}o}\right)^{\frac{1}{2}} \tag{4.129}$$

The diffusion coefficient increases with temperature according to a $T^{\frac{1}{2}}$ law (since it is proportional to the molecular velocity) while the branching rate constant k_b has an activation energy of about 20 kcal mole^{-1}, close to its endothermicity. Equation (4.129) therefore predicts the slow decrease of P_l with increasing temperature that is observed experimentally. The dependence of (4.129) on the reciprocal of the vessel diameter is also in accord with experiment. The effect of added inert gases is to reduce the diffusion coefficient and hence to lower the limit P_l.

At the upper limit P_u, gas-phase termination is more important than surface termination so that equation (4.128) becomes approximately,

$$0 \simeq 2k_{b}oP_u - k_{tg}omP_u^2$$

or,

$$P_u \simeq \frac{2k_b}{k_{tg}m} \tag{4.130}$$

As noted above the branching rate increases with temperature due to its activation energy of about 20 kcal mole^{-1}. The gas-phase termination rate constant k_{tg} has a small negative activation energy of about -2 kcal mole^{-1}. The apparent activation energy predicted from (4.130) for the upper limit is therefore 22 kcal mole^{-1} in good agreement with the experimental value. The absence of surface effects on the upper limit is also explained by (4.130) while the effect of adding inert gases is to increase m and hence to lower the upper limit by assisting the third-order gas-phase termination. Thus all the observed characteristics of the first and second explosion limits can be quantitatively accounted for by the branched chain mechanism outlined above.

It remains to consider the cause of the third limit. As noted earlier the third limit is probably in general a combination of a thermal limit and a branched chain limit. At the higher pressures particularly, the thermal contribution to the explosion is important. The chemical effect is attributed to the limited stability of the HO_2' radical which is formed in the gas-phase termination reaction [4.103]. In order for this termination to be effective this species must live long enough for it to reach the wall of the vessel by diffusion [4.104] and so be removed from the reaction by forming H_2O_2, etc. At high pressures the diffusion of HO_2' to the wall is sufficiently slowed down that the alternative reaction,

$$HO_2 + H_2 \rightarrow H_2O + OH^{\cdot} \qquad [4.106]$$

becomes appreciable and so re-starts the reaction chain. This is equivalent to a decrease in the rate of gas-phase termination so that the critical condition $\phi = 0$ is again reached and explosion occurs.

The same type of mechanism that has been applied to account for the hydrogen–oxygen reaction may also be applied to some other combustion reactions, for example those of carbon monoxide, phosphorus and hydrogen sulphide. We will now consider very briefly some combustion processes which differ somewhat in principle from the case considered above.

4.4d The combustion of hydrocarbons—degenerate branching

These reactions are more complex than the hydrogen–oxygen reaction and we shall outline only briefly some of the principal points of difference. Unlike the H_2/O_2 case the termination reactions in hydrocarbon combustion do not clearly separate and lead to distinct lower and upper isothermal explosion limits. Only one explosion limit is found for hydrocarbon combustions and it is mainly thermal in nature. This single limit varies in rather a different way with temperature and pressure as can be seen by comparing the typical explosion diagram for a hydrocarbon in Figure 4.10 with that of Figure 4.9 for hydrogen. Also more complex phenomena occur outside the explosion limit as illustrated in Figure 4.10. In the cool flame region periodic blue light emission is observed accompanied by pressure pulses and peaks in the concentrations of certain reaction intermediates such as aldehydes. The light emitted is the fluorescence spectrum of electronically excited formaldehyde and the intensity is very weak. Under conditions in the vicinity of the cool flame region there is a range of temperature over which the reaction shows a marked negative temperature coefficient. Outside this range the rate increases very

rapidly with temperature in the normal way. This is illustrated in Figure
4.11 which shows the rate of reaction plotted against temperature for
several stages in a self-accelerating hydrocarbon combustion in this
temperature range. Whereas the induction periods observed in the

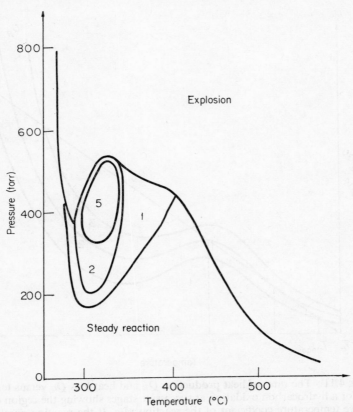

FIGURE 4.10   The explosion and cool-flame limit diagram for a stoichiometric
mixture of propane and oxygen. The numbered regions are the cool-flame
regions, the numbers indicating the number of cool flames observed (Newitt and
Thornes, *J. Chem. Soc.*, 1656 (1937)).

hydrogen–oxygen reaction are usually very small and of the order of
milliseconds, for hydrocarbon combustions they may be as long as several
minutes. This implies that the net branching factor ϕ exceeds zero by only
a very small margin as can be seen from equation (4.123). At such very
small ϕ values clearly the reaction may be over before explosion can occur.
Thus the autocatalytic reaction may start to accelerate but before the rate

becomes high enough to be classed as explosive, depletion of the reactant concentrations causes the rate to start to decline towards zero at completion. Sigmoid extent of reaction versus time curves result and the distinction between slow reaction and explosion is less clear-cut. All these features

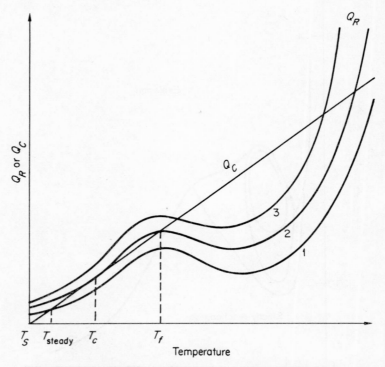

FIGURE 4.11   The rates of heat production Q_R and heat loss Q_C versus temperature for a hydrocarbon oxidation at successive stages showing the region of the negative temperature coefficient of the reaction rate. If the reaction accelerates from curve 1 to curve 2 then the steady gas temperature will rise smoothly from T_{steady} to T_c and then jump suddenly from T_c to T_f. A small drop in reaction rate would then cause the temperature to fall back sharply to T_c.

are attributed to what is called *degenerate branching*. This means that the main chain reaction consuming the hydrocarbon is unbranched but that an intermediate product of the chain, called the *degenerate branching agent*, occasionally undergoes a side reaction which leads to chain branching. In this way very small magnitudes for the net branching factor over a considerable range of conditions can be explained. The degenerate branching agents are believed to be aldehydes and hydroperoxides which are produced in the main chain by straight propagating steps such as,

$$RCH_2CH_2^{\cdot} + O_2 \rightarrow RCH_2CH_2OO^{\cdot} \qquad\qquad [4.107]$$

$$RCH_2CH_2^{\cdot} + O_2 \rightarrow RCH{=}CH_2 + HO_2^{\cdot} \qquad [4.108]$$

$$RCH_2CH_2OO^{\cdot} \rightarrow RCH_2CHO + OH^{\cdot} \qquad\qquad [4.109]$$

$$RCH_2CH_2OO^{\cdot} + RCH_2CH_3 \rightarrow RCH_2CH_2OOH + RCH_2CH_2^{\cdot} \qquad [4.110]$$

$$HO_2^{\cdot} + RCH_2CH_3 \rightarrow H_2O_2 + RCH_2CH_2^{\cdot} \qquad [4.111]$$

$$OH^{\cdot} + RCH_2CH_3 \rightarrow H_2O + RCH_2CH_2^{\cdot} \qquad [4.112]$$

These intermediate products decompose mainly by non-branching steps such as,

$$RCH_2CH_2OOH \rightarrow RCH_2CHO + H_2O \qquad [4.113]$$

$$RCH_2CHO + OH^{\cdot} \rightarrow RCH_2CO^{\cdot} + H_2O \qquad [4.114]$$

$$RCH_2CO^{\cdot} \rightarrow RCH_2^{\cdot} + CO \qquad\qquad [4.115]$$

Occasionally however these reactive intermediates may undergo reactions which cause branching of the reaction chain, for example,

$$RCH_2CHO + O_2 \rightarrow RCH_2CO^{\cdot} + HO_2^{\cdot} \qquad [4.116]$$

$$RCH_2CH_2OOH \rightarrow RCH_2CH_2O^{\cdot} + OH^{\cdot} \qquad [4.117]$$

If the activation energies for the non-branching decompositions of these intermediates are higher than those for the reactions which lead to branching there should be a region of temperature in which increase of temperature causes a reduction in the overall reaction rate because of the decreasing proportion of intermediate that leads to chain branching. In this way the region of the negative temperature coefficient observed experimentally may be understood. Thermal instability caused by this unusual temperature variation of rate can produce sudden jumps in the temperature of the reaction mixture (see Figure 4.11) at certain points during the reaction. It is believed that these sudden temperature changes are the source of the phenomenon of cool flames. A satisfactory detailed explanation of the cause of these interesting oscillatory phenomena is still to be found. Likewise many other details of hydrocarbon combustion mechanisms are still obscure. The reaction steps outlined above are intended to be illustrative examples only of the types of reactions believed to be responsible. The basic principles of degenerate branching outlined above are probably correct, though the actual steps used to exemplify these principles may not be.

8

SUGGESTIONS FOR FURTHER READING

Textbooks

P. G. Ashmore, F. S. Dainton and T. M. Sugden, *Photochemistry and Reaction Kinetics*, Cambridge University Press, Cambridge, 1967.

S. W. Benson, *Foundations of Chemical Kinetics*, McGraw-Hill, New York, 1960.

F. S. Dainton, *Chain Reactions*, Methuen, London, 1956.

E. W. R. Steacie, *Atomic and Free Radical Reactions*, Vol. I and II, Reinhold, New York, 1954.

Reviews

Advan. Photochem., **2**, 1 (1964); *Advan. Photochem.*, **4**, 1 (1966); *Ann. Rev. Phys. Chem.*, **16**, 397 (1965); *Ann. Rev. Phys. Chem.*, **17**, 173 (1966); *Ann. Rev. Phys. Chem.*, **18**, 261 (1967); *Progr. Reaction Kinetics*, **1**, 41 (1963); *Progr. Reaction Kinetics*, **3**, 171 (1965); *Quart. Rev.*, **12**, 61 (1958).

CHAPTER 4 PROBLEMS

1. A mixture of hydrogen and bromine was irradiated with light of wavelength 5000Å. The rate of light absorption was $0 \cdot 127$ watts. After an exposure of 1000 seconds analysis showed that 5×10^{-6} moles of hydrogen bromide had been formed. What is the quantum yield for the formation of HBr? Assuming that the quantum yield of the primary dissociation is unity for the conditions used in this experiment calculate the average chain length.

2. Calculate the induction period $t_{0.9}$ (i.e. the time taken to reach a rate of 90% of the steady state rate) for the thermal H_2/Br_2 reaction at 500°K when both reactants have partial pressures of 100 torr. (Use the rate constant data on p. 179 and assume rapid establishment of a quasi-stationary state). What percentage of reaction occurs during this period? What is the average lifetime of a chain centre in the steady state?

3. Calculate the chain length in the pyrolysis of 500 torr of ethane at 600°C. (Use the rate constants given on pp. 186, 187.)

4. The following mechanism has been proposed for the pyrolysis of acetaldehyde,

$$CH_3CHO \xrightarrow{1} CH_3^{\cdot} + CHO^{\cdot}$$

$$CHO^{\cdot} \xrightarrow{2} CO + H$$

$$H + CH_3CHO \xrightarrow{3} H_2 + CH_3CO^{\cdot}$$

$$CH_3CO^{\cdot} \xrightarrow{4} CH_3^{\cdot} + CO$$

$$CH_3^{\cdot} + CH_3CHO \xrightarrow{5} CH_4 + CH_3CO^{\cdot}$$

$$2CH_3^{\cdot} \xrightarrow{6} C_2H_6$$

(a) Derive an expression for the rate of formation of CH_4 in the steady state in terms of the rate constants of the above elementary reactions and the concentration of reactant.

(b) Calculate the order of reaction under conditions such that all the unimolecular steps above are first order.

(c) Assuming the mechanism remains valid as the pressure is reduced how will the order of reaction change?

(d) Given that step (1) has an activation energy of 76 kcal mole^{-1} use the data in Tables 4.1 and 4.2 to calculate the overall activation energy for the pyrolysis of acetaldehyde at high pressures.

5. The following mechanism has been proposed for the pyrolysis of acetone,

$$CH_3COCH_3 \xrightarrow{1} CH_3CO^\cdot + CH_3^\cdot$$

$$CH_3CO^\cdot \xrightarrow{2} CH_3^\cdot + CO$$

$$CH_3^\cdot + CH_3COCH_3 \xrightarrow{3} CH_4 + CH_3COCH_2^\cdot$$

$$CH_3COCH_2^\cdot \xrightarrow{4} CH_3^\cdot + CH_2{=}CO$$

$$CH_3^\cdot + CH_3COCH_2^\cdot \xrightarrow{5} CH_3COC_2H_5$$

Assuming the chains are long apply the steady state method to obtain an expression for the rate of reaction and hence deduce the order of reaction at high pressures.

6. The polymerization of 10^{-6} mole cc^{-1} of ethylene at 100°C was initiated by 10^{-8} mole cc^{-1} of di-t-butyl peroxide which decomposes,

$$((CH_3)_3CO)_2 \xrightarrow{1} 2(CH_3)_3CO^\cdot$$

$$(CH_3)_3CO^\cdot \xrightarrow{2} CH_3COCH_3 + CH_3^\cdot$$

with a first-order rate constant $k_1 = 10^{15} \exp(-37,000/RT)$ sec^{-1}. The initial rate of polymerization was measured as

$$-d[C_2H_4]/dt = 2 \cdot 0 \times 10^{-13} \text{ mole cc}^{-1} \text{ sec}^{-1}$$

Neglecting chain transfer processes estimate the molecular weight of the polymer assuming that termination occurs exclusively by, (a) combination, (b) disproportionation.

The rotating sector method was used to estimate the average life-time τ of a growing polymer radical under conditions similar to those of the thermal 'catalysed' reaction above as 5·0 sec. Calculate the concentration of active centres in the thermal reaction system. Hence calculate the rate constants k_t and k_p for the termination and propagation reactions.

7. The thermal decomposition of azomethane has a first-order rate constant, $k = 10^{15 \cdot 67} \exp(-51,200/RT)$ sec^{-1} and is 43 kcal mole^{-1} exothermic. Experiments to study the thermal explosion limit were conducted in a spherical vessel of surface to volume ratio $S/V = 1$ cm^{-1}. The rate of heat loss from the gas, Q_c, can be approximated by Newton's law of cooling, $Q_c = Sh(T - T_s)$, where T is the mean temperature of the gas, T_s is the temperature of the vessel, S is the surface area of the vessel, and h the heat transfer

coefficient had a value of $8 \cdot 18 \times 10^{-5}\,\mathrm{cal\,cm^{-2}\,deg^{-1}\,sec^{-1}}$ under the conditions of these experiments. Calculate the mean gas temperature just below the explosion limit when the vessel temperature is 360°C and hence calculate the pressure of azomethane (in torr) at the explosion limit. What would be the effect on this pressure of (a) reducing the diameter of the vessel? (b) diluting the reactant with helium?

8. The rate of reaction of a mixture of hydrogen and oxygen was measured by a fast recording pressure transducer during the induction period that preceeds explosion. Some results were,

Time (sec)	2	3	4	5	6
Rate (arbitrary units)	1·60	4·78	13·4	36·8	100·6

Assuming that the reaction mixture remained approximately isothermal during these observations evaluate the net branching factor ϕ. (Assume that $t > 1/\phi$ for these results).

9. The second explosion limit of stoichiometric mixtures of hydrogen and oxygen was found to vary with temperature as follows.

T (°C)	450	500	550
P_u (torr)	20	50	110

Given that the termination reaction $H + O_2 + M \rightarrow HO_2^{\cdot} + M$ has an activation energy of $-1 \cdot 6\,\mathrm{kcal\,mole^{-1}}$ calculate from these results the activation energy of the branching reaction, $H + O_2 \rightarrow OH^{\cdot} + O$. Use this value to predict how the first (lower) explosion limit of the reaction should vary with temperature.

10. The second explosion limit of mixtures of hydrogen and oxygen was found to vary with the stoichiometry of the mixture as follows,

Total pressure at explosion limit (torr)	89	92·5	96	100	104
(P_{H_2}/P_{O_2}) in mixture	0·780	0·542	0·371	0·250	0·156

Use these results to estimate the relative efficiencies of H_2 and O_2 as third bodies in the termination reaction $H + O_2 + M \rightarrow HO_2^{\cdot} + M$.

APPENDIX A

THE CALCULATION OF EQUILIBRIUM CONSTANTS OF GASEOUS REACTIONS BY STATISTICAL MECHANICS (A SIMPLE NON-RIGOROUS TREATMENT)

For an ideal gas we can suppose that each molecule possesses a set of energy levels as depicted in Figure A.1. These energy levels are very closely spaced but may be divided into groups of equal energy, there being

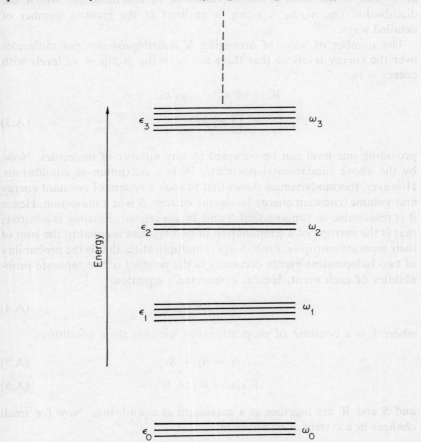

FIGURE A.1   The energy levels available to a molecule of an ideal gas (arranged in groups of equal energy).

w_i levels in the group with energy ε_i. Let the number of gas molecules that have an energy ε_i (in the group of w_i levels) be n_i, then,

$$\text{the total number of molecules, } N = \sum_{i=0}^{\infty} n_i \tag{A.1}$$

$$\text{the total energy, } E = \sum_{i=0}^{\infty} n_i \varepsilon_i \tag{A.2}$$

Since the gas is ideal there is no communally shared energy.

The fundamental postulate of statistical mechanics is that for given values of E and N the most probable values of n_i are those for which the distribution $(n_0, n_1, n_2 \ldots)$ can be realized in the greatest number of detailed ways.

The number of ways of arranging N *indistinguishable* gas molecules over the energy levels so that there are n_i in the group of w_i levels with energy ε_i is,

$$W = w_0^{n_0} w_1^{n_1} \ldots / n_0! n_1! \ldots$$

$$= \prod_{i=0}^{\infty} w_i^{n_i} / n_i! \tag{A.3}$$

providing one level can be occupied by any number of molecules. Now, by the above fundamental postulate, W is a maximum at equilibrium. However, thermodynamics shows that in such a system of constant energy and volume (constant energy levels) the entropy S is at a maximum. Hence it is reasonable to suppose that S and W are related. Entropy is additive, that is the entropy of a combination of two systems is equal to the sum of their separate entropies. Probability is multiplicative, that is the probability of two independent events occurring is the product of the separate probabilities of each event. Hence, Boltzmann's equation,

$$S = k \ln W \tag{A.4}$$

where k is a constant of proportionality, satisfies these conditions,

$$S_{1+2} = S_1 + S_2 \tag{A.5}$$

$$W_{1+2} = W_1 \times W_2 \tag{A.6}$$

and S and W are together at a maximum at equilibrium. Now for small changes in a system at equilibrium

$$dE = T\,dS - P\,dV + \left(\frac{\partial E}{\partial N}\right)_{S,V} dN \tag{A.7}$$

where the thermodynamic symbols have their usual meaning. Writing,

$$\mu = \left(\frac{\partial E}{\partial N}\right)_{S,V} = \left(\frac{\partial A}{\partial N}\right)_{T,V} = \left(\frac{\partial G}{\partial N}\right)_{T,P} \tag{A.8}$$

for the chemical potential, then at constant volume from (A.7)

$$dS = \frac{dE - \mu \, dN}{T} \tag{A.9}$$

hence from (A.4), $kd(\ln W) = (dE - \mu \, dN)/T$

$$\therefore \quad k \sum_{i=0}^{\infty} \frac{\partial(\ln W)}{\partial n_i} \, dn_i = \frac{1}{T} \sum_{i=0}^{\infty} \frac{\partial E}{\partial n_i} \, dn_i - \frac{\mu}{T} \sum_{i=0}^{\infty} \frac{\partial N}{\partial n_i} \, dn_i$$

$$= \frac{1}{T} \sum_{i=0}^{\infty} \varepsilon_i \, dn_i - \frac{\mu}{T} \sum_{i=0}^{\infty} dn_i \tag{A.10}$$

This is true for all values of dn_i, hence setting $dn_i = 1$ and $dn_j = 0$ for all $j \neq i$,

$$k \frac{\partial(\ln W)}{\partial n_i} = \frac{\varepsilon_i - \mu}{T} \quad \text{for all } i \tag{A.11}$$

Now from (A.3)

$$\ln W = \sum_{i=0}^{\infty} n_i \ln w_i - \sum_{i=0}^{\infty} \ln(n_i!) \tag{A.12}$$

For large values of n_i, Stirling's approximation

$$\ln(n!) = n \ln n - n \tag{A.13}$$

may be used, therefore

$$\ln W = \sum_{i=0}^{\infty} n_i(\ln w_i - \ln n_i + 1) \tag{A.14}$$

therefore

$$\frac{\partial \ln W}{\partial n_i} = \ln w_i - \ln n_i \tag{A.15}$$

Substituting into (A.11) gives,

$$\ln w_i - \ln n_i = \frac{\varepsilon_i - \mu}{kT} \tag{A.16}$$

or,

$$n_i = w_i \exp\left(\frac{\mu - \varepsilon_i}{kT}\right) \tag{A.17}$$

Now

$$N = \sum_{i=0}^{\infty} n_i \tag{A.1}$$

$$= e^{\mu/kT} \sum_{i=0}^{\infty} w_i \, e^{-\varepsilon_i/kT} \tag{A.18}$$

Defining

$$f = \sum_{i=0}^{\infty} w_i \, e^{-\varepsilon_i/kT} \tag{A.19}$$

we get,

$$N = f \, e^{\mu/kT} \tag{A.20}$$

therefore

$$n_i/N = \frac{w_i \, e^{-\varepsilon_i/kT}}{f} \tag{A.21}$$

f is called the partition function and is a function of temperature and volume (energy levels) only. It is independent of the n_i's. From (A.20)

$$\mu = kT \ln\left(\frac{N}{f}\right) \tag{A.22}$$

Now from (A.8)

$$A = \int_0^N \mu \, dN \quad \text{at constant } T,V \tag{A.23}$$

Using (A.22)

$$A = \int_0^N kT \, \ln\left(\frac{N}{f}\right) dN \tag{A.24}$$

$$= -NkT \ln f + kT(N \ln N - N)$$

or,

$$A = -kT \ln(f^N/N!) \tag{A.25}$$

Hence all other thermodynamic quantities may be related to f. The gas pressure P is defined by

$$P = -\left(\frac{\partial A}{\partial V}\right)_T \tag{A.26}$$

hence

$$P = kT \frac{\partial \ln(f^N/N!)}{\partial V} \tag{A.27}$$

For molecules in their ground electronic state, to a good approximation

$$\varepsilon = \varepsilon_t + \varepsilon_r + \varepsilon_v \tag{A.28}$$

where the suffixes, t, r, v, stand for the translational, rotational and vibrational contributions to the total energy. Substituting (A.28) into (A.19) gives

$$f = f_t f_r f_v \qquad (A.29)$$

where

$$f_t = \sum_{i=0}^{\infty} (w_i)_t \, e^{-(\varepsilon_i)t/kT} \qquad \text{etc.} \qquad (A.30)$$

and are called the translational, rotational and vibrational partition functions respectively. For free one-dimensional translational motion in a length l the energy levels are,

$$\varepsilon = \frac{n^2 h^2}{8ml^2} \qquad (A.31)$$

where n is the translational quantum number in one dimension and m is the mass of the molecule. Hence for 3 dimensions

$$f_t = \left[\sum_{n=1}^{\infty} \exp\left(\frac{-n^2 h^2}{8ml^2 kT} \right) \right]^3 \qquad (A.32)$$

which to a good approximation at normal temperatures is

$$f_t = \left[\int_0^{\infty} \exp\left(\frac{-n^2 h^2}{8ml^2 kT} \right) dn \right]^3 \qquad (A.33)$$

$$= V \left(\frac{2\pi mkT}{h^2} \right)^{\frac{3}{2}} \qquad (A.34)$$

where $V = l^3$ is the volume of the container.

For a heteronuclear diatomic molecule of moment of inertia, I, the rotational energy levels are

$$\varepsilon_J = \frac{J(J+1)h^2}{8\pi^2 I} \qquad (A.35)$$

(measuring ε_J from the lowest level, $J = 0$, as zero), and have $(2J+1)$ degeneracy. Hence the rotational partition function is,

$$f_r = \sum_{J=0}^{\infty} (2J+1) \exp\left(\frac{-J(J+1)h^2}{8\pi^2 IkT} \right) \qquad (A.36)$$

which, to a good approximation at normal T, is

$$f_r = \int_0^{\infty} (2J+1) \exp\left(\frac{-J(J+1)h^2}{8\pi^2 IkT} \right) dJ \qquad (A.37)$$

$$= \frac{8\pi^2 IkT}{h^2} \qquad (A.38)$$

For a homonuclear diatomic, as explained on p. 136, this expression must be divided by the symmetry number $\sigma = 2$. For a linear polyatomic molecule this same result applies with $\sigma = 2$ if the molecule has a plane of symmetry and $\sigma = 1$ if it has not. For a non-linear polyatomic molecule having no symmetry and principal moments of inertia I_A, I_B, and I_C, the corresponding result is

$$f_r = \left(\frac{8\pi^2(\pi I_A I_B I_C)^{\frac{1}{2}}kT}{h^2}\right)^{\frac{3}{2}} \tag{A.39}$$

Again for a molecule with some symmetry this result must be divided by the symmetry number σ which is the number of indistinguishable positions into which the molecule can be turned by rotations. For a simple harmonic oscillator of frequency ν the energy levels relative to the lowest (zero point) level are

$$\varepsilon_n = nh\nu \tag{A.40}$$

hence,

$$f_v = \sum_{n=0}^{\infty} \exp\left(\frac{-nh\nu}{kT}\right)$$

$$= (1 - e^{-h\nu/kT})^{-1} \tag{A.41}$$

for a single oscillator. For a molecule with several (n) vibrational modes of frequencies ν_i,

$$f_v = \prod_{i=1}^{n} (1 - e^{-h\nu_i/kT})^{-1} \tag{A.42}$$

From equations, (A.34), (A.38), (A.39), (A.41) and (A.42) the partition function for any gaseous molecules can be calculated provided the mass, moments of inertia and vibration frequencies are known. Hence from (A.25) the (Helmholtz) free energy and therefore all other thermodynamic quantities may be calculated in terms of the known partition function f. In particular equation (A.27) for the pressure gives

$$P = kT\frac{\partial \ln V^N}{\partial V}$$

since only f_t depends on V [equation (A.34)] hence

$$P = \frac{NkT}{V} \tag{A.43}$$

which shows that

$$k = \frac{R}{N} \qquad (A.44)$$

where R is the ideal gas constant in the equation

$$PV = RT \qquad (A.45)$$

The Boltzmann constant k is therefore equal to the gas constant per molecule.

To apply these results to the calculation of equilibrium constants, consider the equilibrium reaction whose stoichiometric equation is (see [1.6])

$$0 = \sum_{i=1}^{m} \nu_i X_i \qquad [A.1]$$

At equilibrium

$$(dA)_{T,V} = 0 \qquad (A.46)$$

Hence from (A.8)

$$\sum_{i=1}^{m} \mu_i \, dN_i = 0 \qquad (A.47)$$

The dN_i are related by the stoichiometric coefficients, ν_i, hence

$$\sum_{i=1}^{m} \nu_i \mu_i = 0 \qquad (A.48)$$

Substituting from equation (A.22) gives

$$kT \sum_{i=1}^{m} \nu_i \ln (N_i/f_i) = 0 \qquad (A.49)$$

$$\therefore \qquad \prod_{i=1}^{m} (N_i/f_i)^{\nu_i} = 1 \qquad (A.50)$$

i.e.

$$\prod_{i=1}^{m} N_i^{\nu_i} = \prod_{i=1}^{m} f_i^{\nu_i} \qquad (A.51)$$

Dividing each term of the products by the volume V to convert N to the molecular concentration (N/V)

$$\prod_{i=1}^{m} \left(\frac{N_i}{V}\right)^{\nu_i} = \prod_{i=1}^{m} \left(\frac{f_i}{V}\right)^{\nu_i} \qquad (A.52)$$

By definition

$$\prod_{i=1}^{m} \left(\frac{N_i}{V}\right)^{\nu_i} = K_c \qquad (A.53)$$

where K_c is the equilibrium constant in terms of molecular concentrations hence, writing

$$Q' = \frac{f}{V} \tag{A.54}$$

where Q' is the partition function per unit volume [see (A.34)] we obtain,

$$K_c = \prod_{i=1}^{m} Q_i'^{\nu_i} \tag{A.55}$$

which is a function of temperature only.

This formula enables any equilibrium constant for reactions between ideal gases to be calculated from the equations for the partition function. In calculating the partition functions to be used with (A.55) a fixed energy zero must of course be used throughout for all the different chemical substances involved in the reaction. It is usually convenient when evaluating the partition function for any particular substance to choose the energy zero as equal to the energy of the lowest level for that substance (i.e. to set $\varepsilon_0 = 0$) as was done when deriving (A.38), (A.39), (A.41), (A.42). Calling this partition function per unit volume Q we have from (A.19) and (A.54)

$$Q = \frac{1}{V} \sum_{i=0}^{\infty} w_i \, e^{-(\varepsilon_i - \varepsilon_0)/kT} \tag{A.56}$$

or,

$$Q = Q' \, e^{\varepsilon_0/kT} \tag{A.57}$$

where ε_0 is measured from any arbitrary zero. Substituting into (A.55), for Q' from (A.57)

$$K_c = \prod_{i=1}^{m} Q_i^{\nu_i} \, e^{-\nu_i(\varepsilon_0)_i/kT} \tag{A.58}$$

$$K_c = e^{-\sum_{i=1}^{m} \nu_i(\varepsilon_0)_i/kT} \prod_{i=1}^{m} Q_i^{\nu_i} \tag{A.59}$$

Now,

$$\sum_{i=1}^{m} \nu_i(\varepsilon_0)_i = \Delta\varepsilon_0 \tag{A.60}$$

which is the difference in the lowest energy levels of products and reactants i.e. it is the energy change in the reaction, between the number of molecules as written in the stoichiometric equation, at $0°K$.

Using molar quantities rather than molecular ones

$$\frac{\Delta\varepsilon_0}{kT} = \frac{\Delta E_0}{RT} \tag{A.61}$$

we get,

$$K_c = \prod_{i=1}^{m} Q_i^{\nu_i}\, e^{-\Delta E_0/RT} \tag{A.62}$$

which is the required relation between the molecular properties and the equilibrium constant.

APPENDIX B

CHAPTER 1 ANSWERS

1. (a) $m = 1$, $r = -\dfrac{d[N_2O]}{dt} = \dfrac{d[N_2]}{dt} = \dfrac{d[O]}{dt}$

    (b) $m = 2$, $r = -\dfrac{d[C_2H_4]}{dt} = -\dfrac{d[HCl]}{dt} = \dfrac{d[C_2H_5Cl]}{dt}$

    (c) Cannot be elementary, $r = -\dfrac{d[C_5H_{12}]}{dt} = -\dfrac{1}{8}\dfrac{d[O_2]}{dt} = \dfrac{1}{5}\dfrac{d[CO_2]}{dt}$
    $= \dfrac{1}{6}\dfrac{d[H_2O]}{dt}$

    (d) $m = 2$, $r = -\dfrac{d[H_2]}{dt} = -\dfrac{d[F_2]}{dt} = \dfrac{1}{2}\dfrac{d[HF]}{dt}$

    (e) $m = 3$, $r = -\dfrac{1}{2}\dfrac{d[H]}{dt} = \dfrac{d[H_2]}{dt}$

    (f) Cannot be elementary, (consider reverse), $r = -\dfrac{d[Pb(C_2H_5)_4]}{dt}$
    $= \dfrac{d[Pb]}{dt} = \dfrac{1}{4}\dfrac{d[C_2H_5^{\cdot}]}{dt}$

    (g) Cannot be elementary, $r = -\dfrac{1}{2}\dfrac{d[H]}{dt} = -\dfrac{1}{2}\dfrac{d[O]}{dt} = \dfrac{d[H_2O_2]}{dt}$

    (h) $m = 2$, $r = -\dfrac{d[Na]}{dt} = -\dfrac{d[C_2H_5Cl]}{dt} = \dfrac{d[NaCl]}{dt} = \dfrac{d[C_2H_5^{\cdot}]}{dt}$

2. Partial order w.r.t. A = 0·5
   Partial order w.r.t. B = 1·5
   Total order = 2·0
   $_2k = 1·0 \times 10^{-5}$ torr^{-1} sec^{-1} = $4·2 \times 10^2$ cc mole^{-1} sec^{-1}

3. At 1·0 torr order = 2·0
   at 100 torr order = 1·0

4. Total order = 1·25; k cannot be calculated from the data given

5. $x_A = 1·2$, $x_B = 0·6$

6. Total order = 2·50

7. $x = 2·0$, $_2k = 2·27 \times 10^3$ cc mole^{-1} sec^{-1}

8. $_1k = 3·325 \times 10^{-4}$ sec^{-1}, true order = 2·0, $_2k = 1·68 \times 10^3$ cc mole^{-1} sec^{-1}, $E_a = 28$ kcal mole^{-1}, $A_2 = 4·5 \times 10^{16}$ cc mole^{-1} sec^{-1}

9. $E_a = 28·9$ kcal mole^{-1}, $_2A = 10^{11·5}$ cc mole^{-1} sec^{-1}

10. $E_a = 45·5$ kcal mole^{-1}

11. $n = -2·0$

12. $x = 1.30$

13. For reverse $_2k = 10^{13.58} \exp(-660/RT)$ cc mole^{-1} sec^{-1}

14. 33.80 kcal mole^{-1} $(E_a = \Delta H^\circ - \Delta \nu RT)$

15. $_2k = k_1k_3/(k_2 + k_3)$. Arrhenius plot curved, E_a changes from 10 kcal mole^{-1} to zero as the temperature increases.

16. Rate $= k_4 \left(\dfrac{k_1k_3}{k_2k_4}[A][B]^2\right)^{2/3}$

$x_A = 0.667;\ x_B = 1.333;\ \Sigma x = 2.00$

CHAPTER 3 ANSWERS

1. $_2r = 1.83 \times 10^{28}$ collisions cc^{-1} sec^{-1}, $t_{1/2} = 5.2 \times 10^{-10}$ sec

2. $Z = 2kT/\pi\eta = 2.79 \times 10^{-10}$ cc molecule^{-1} sec^{-1} = 1.68×10^{14} cc mole^{-1} sec^{-1}

3. 6.3×10^{-15}

4. 1.9×10^{-3} $(n = 18)$

5. $_2A = e^{1/2}Z = 8.54 \times 10^{13}$ cc mole^{-1} sec^{-1}

6. $P = A/e^{1/2}Z = 9.23 \times 10^{-4}$

7. At $1000°K$, $k = 6.35 \times 10^{14} \exp(-9,000/RT)$ cc mole^{-1} sec^{-1}
   At $3000°K$, $k = 1.1 \times 10^{15} \exp(-11,000/RT)$ cc mole^{-1} sec^{-1}

8. $_3r = 5.41 \times 10^{26}$ collisions cc^{-1} sec^{-1}, $t_{1/2} = 6.85 \times 10^{-8}$ sec

9. $E_a = -4.7 + \frac{1}{2}RT = -4.4$ kcal mole^{-1}

10. $P = \left(\dfrac{E-50}{E}\right)^9,\ E \geqslant 50$

11. $NO_2Cl \rightarrow NO_2 + Cl$
    $Cl + NO_2Cl \rightarrow NO_2 + Cl_2$
    Overall rate equals that of the first step, the unimolecular decomposition of NO_2Cl which is in its second-order region, hence the difference between the true and time-orders. $_2k^0 = k_a$. Hence get n by successive approximations $2n = 6.6$

12. $_{1/2}P = 940$ torr

13. $_2E^0_{af} = 29.6$ kcal mole^{-1}, $_2E^0_{ab} = -8.5$ kcal mole^{-1}

14. $k_a \simeq 8 \times 10^7$ cc mole^{-1} sec^{-1}

15. $\Delta V^\ddagger = 6.35$ kcal mole^{-1}

16. (a) $S = 2$, $\sigma_A\sigma_B/\sigma^\ddagger = 1$
    (b) $S = 1$, $\sigma_A\sigma_B/\sigma^\ddagger = 1$
    (c) $S = 1$, $\sigma_A\sigma_B/\sigma^\ddagger = 1/2$
    (d) $S = 4$, $\sigma_A\sigma_B/\sigma^\ddagger = 4$
    (e) $S = 12$, $\sigma_A/\sigma^\ddagger = 6$
    (f) $S = 8$, $\sigma_A/\sigma^\ddagger = 2$

17. $(\Delta E^\ddagger)^\circ_c = 17.4$ kcal mole^{-1}, $(\Delta H^\ddagger)^\circ_c = 16.4$ kcal mole^{-1}, $(\Delta S^\ddagger)^\circ_c = 0.0617$ cal deg^{-1} mole^{-1}, $(\Delta G^\ddagger)^\circ_c = 16.37$ kcal mole^{-1}

18. $k = 9.9 \times 10^{13} \exp(-5,570/RT) = 6.0 \times 10^{12}$ cc mole^{-1} sec^{-1}

CHAPTER 4 ANSWERS

1. $\Phi = 9\cdot44 \times 10^{-3}$, $n = 2\cdot38 \times 10^{-3}$
2. $t_{0,9} = 115$ sec, $\tau = 76\cdot7$ sec, amount of reaction $= 6\cdot2 \times 10^{-4}\%$
3. $n = 25$
4. (a) $\dfrac{d[CH_4]}{dt} = k_5 \left(\dfrac{k_1}{k_6}\right)^{\frac{1}{2}} [CH_3CHO]^{\frac{3}{2}}$

    (b) $\frac{3}{2}$

    (c) k_1 and k_6 become pressure dependent at different pressures. At very low pressures order still $\frac{3}{2}$.

    (d) $E_a = E_5 + \frac{1}{2}(E_1 - E_6) = 7\cdot8 + \frac{1}{2} \times 76 = 45\cdot8$ kcal mole^{-1}
5. Rate $= \left(\dfrac{k_1 k_3 k_4}{k_5}\right)^{\frac{1}{2}} [CH_3COCH_3]$, first-order
6. (a) M.W. $\simeq 2{,}800$, (b) M.W. $\simeq 1{,}400$, $\Sigma[R^\cdot] = 2 \times 10^{-14}$ mole cc^{-1}, $k_t = 10^{13}$ cc mole^{-1} sec^{-1}, $k_p = 10^7$ cc mole^{-1} sec^{-1}
7. $T = 376\cdot4°C$, $P = 55$ torr, (a) increases P, (b) increases P
8. $\phi = 1\cdot0$ sec^{-1}
9. $E_b = 19\cdot3$ kcal mole^{-1}, $P_l \propto T^{1/4} \exp(9{,}650/RT)$
10. $\dfrac{k_{O_2}}{k_{H_2}} = 0\cdot63$

APPENDIX C

PHYSICAL CONSTANTS AND CONVERSION FACTORS

Electronic charge (e)	$4 \cdot 8029 \times 10^{-10}$ e.s.u.
	$1 \cdot 6021 \times 10^{-19}$ coulomb
Planck's constant (h)	$6 \cdot 6256 \times 10^{-27}$ erg sec
Speed of light	$2 \cdot 997925 \times 10^{10}$ cm sec^{-1}
Electron rest mass	$9 \cdot 1090 \times 10^{-28}$ g
Proton rest mass	$1 \cdot 67252 \times 10^{-24}$ g
Neutron rest mass	$1 \cdot 67482 \times 10^{-24}$ g
Avogadro's number (N_{Av})	$6 \cdot 0225 \times 10^{23}$ mole^{-1}
Boltzmann's constant (k)	$1 \cdot 3805 \times 10^{-16}$ erg °K^{-1}
Gas constant (R)	$8 \cdot 31432$ joule °K^{-1} mole^{-1}
	$1 \cdot 9872$ cal. °K^{-1} mole^{-1}
	$0 \cdot 082054$ litre atmosphere °K^{-1} mole^{-1}
	62361 cc torr °K^{-1} mole^{-1}
π	$3 \cdot 14159$
e	$2 \cdot 7183$

1 erg $= 10^{-7}$ joule $= 2 \cdot 3901 \times 10^{-8}$ cal

1 electron volt (eV) $= 1 \cdot 60209 \times 10^{-12}$ erg $= 23 \cdot 053$ kcal mole^{-1}

760 torr $= 1$ atmosphere $= 1 \cdot 01325 \times 10^6$ dyne cm^{-2}

0°C $= 273 \cdot 16$°K

Wavelength equivalent to 1 eV $= 12,398$Å

$\ln X = 2 \cdot 3026 \log_{10} X$

INDEX

Symbols are listed with the page numbers of their first appearances with particular meanings. Greek letter symbols are listed according to their English spelling.

232

DATE DUE

MAY 2 1977			